THE WONDERS OF LIGHT

An Exciting New Viewpoint

by Robert Crane

ISBN # 978-1-4276-2901-2

On the internet at http://www.thewondersoflight.com

All photographs and figures by Robert Crane
Cover photograph descriptions, see Appendix A
Layout design and editing by Plum Crane

Printed in the United States
by Southbury Printing Center, Inc
Southbury, CT

This book is dedicated to Lt. Duncan Mc Laren Crane USMCR 1922-1945

Table of Contents

LIST OF EXPERIMENTS

*In which a new viewpoint for propagating elecromagnetic energy is proposed.

PROLOGUE

Approximately five years ago I had fully recovered from a job fraught with achievement milestones, financial accountability and a boss who by his own admission "ruled by fear." I was enjoying a life of random hobbies, wandering from one subject to another. On the occasion of a grandson's wedding the family gathered at a lodge in the shadow of Mt. Hood. My son and I were chatting over my cup of coffee when he struck. "Dad, you've been working with light all of your life, what is light?"

My studies and experiments over the subsequent five years have been devoted to searching for an answer to that question. I have learned much, but not enough. Experiments have continued to raise more questions than they answer. I have learned that generation and absorption, or detection, of light are well understood, and can be found in many books; but propagation of light is not well understood. In fact, in many respects current beliefs are incorrect. This narrowed the scope of studies and experiments to learning more about the propagation of light. It's the unknowns that are fun to seek.

Early experiments investigated the propagation of radio frequency portions of the electromagnetic spectrum. Among those things learned were:

1. Radio beams consist of many bundles of oscillating fields: electric, E and magnetic, H; with too many to be counted.

2. E and H fields propagate independent of each other and do not always have the same velocity.

3. Maxwell's equations and Planck's quantum "constraint" apply during emission and absorption, but they do not apply to fields propagating in electron-free space. In the absence of Planck's constraint, the size of an E or H field is not bounded.

> *These three points define what I call a "two-field viewpoint" for propagating light.*

Acceptance of this two-field point of view eliminates the quandary posed by the single-photon version of Young's two-slit interferometer.

Recent experiments have involved the propagation of light wavelength beams. Among those things learned were:

1. Our eyes do not respond to the H field portion of light.

2. E and H fields may be separated.

3. The H field portion of a light beam will not be scintillated by atmospheric effects such as those that cause air turbulence.

Ongoing light wavelength experiments include:

1. Improving our understanding of the detection of H field beams by use of Maxwell's "curl."

2. Understanding Fraunhoffer's diffraction as it affects the resolution of an optical telescope.

Looking to the future our curiosity leads us "pretty far out", to wonder if we can create:

1. An X-Ray type camera responding to H field portions of light only.

2. An imaging camera responding to the H field portion of light reflected from scenes.

Come join the curious!

THE WONDERS OF LIGHT

An Exciting New Viewpoint

INTRODUCTION

The following report describes a search for an understanding of light that goes beyond that found in classical physics text books of today. This search covered emission of electromagnetic waves, propagating radio and light fields and then back to electron current at detection. Finding a new viewpoint for understanding electric and magnetic fields was a necessary part of this search.

An experimental approach was selected to resolve most of the questions that came up. Implicit in these experiments was the assumption that light wavelength and radio wavelength energy are the same stuff, differing only in the magnitude of wavelength and/or frequency.

In this book the word "light" will be used to embrace all wavelengths. Experiments covered three different portions of the electromagnetic spectrum.

The intent of this book is to stir the interest of college students in the study of physics; to show that learning Natures's Laws of physics can be fun, does not require extensive understanding of higher math, and that original work can be done. The emphasis is on those features of light which are not known, or in some cases where the known features are described incorrectly in contemporary textbooks. This book relies only upon "classical" physics to describe "why" experimental results are obtained. It studiously avoids contact with theories of quantum electrodynamics developed after 1920. Instead it depends upon experimental

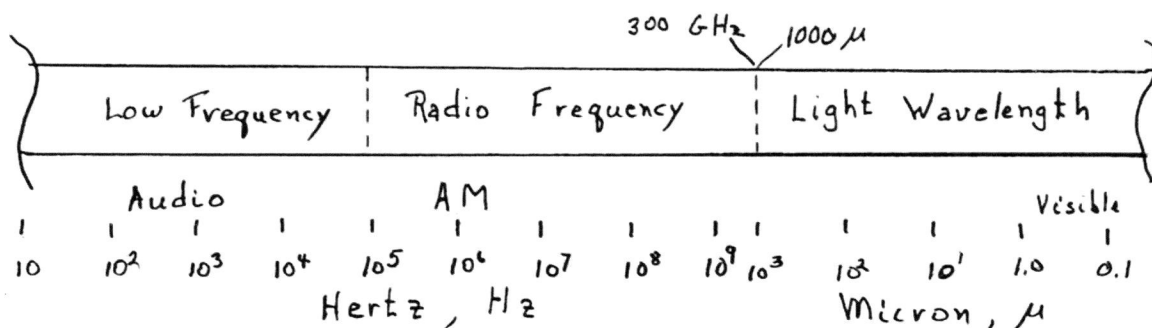

results to support the author's ideas concerning Nature's Laws governing the behavior of light.

Books studied during this project included: "Lectures on Physics", Volume 11, R. Feynman (ref 2), "Physics for Scientists and Engineers", R. A. Serway (ref 3), Electricity and Magnetism, Ralph Winch (ref 4), J. Newman's biography of J. C. Maxwell (ref 5), Mc Graw-Hill Encyclopedia of Science and Engineering (ref 6), "Optics" second edition by E. Hecht (ref 7), "Procedures in Experimental Physics", John Strong (ref 8) and Handbook of Physics (ref 9).

This book has benefitted immeasurably from frequent discussions with my son Duncan Crane. Howard Harris helped with field experiments. My granddaughter Plum Crane converted my engineering writing to readable text, and most importantly she performed the pre-press work of organizing a printable document, without which this book would never have been born. And finally, my wife's sacrifice of a major share of our married life and a large portion of our home must be appreciated and honored.

Fig. 1

1. ABOUT FIELDS

1.1 What is a Field ?

A trail leading to an understanding of light begins with an understanding of electric and magnetic fields. That is not easy because they are weird things. One can not see, hear, touch, taste or smell them. They pass through many objects, including the human body. They have no mass and therefore are immune to Newton's laws of motion and Einstein's gravity, except under extreme conditions. The idea of an electromagnetic field is just not like anything familiar to our brain. They are real but not understood; they are certainly one of the wonders of light. The following is intended to create a "feel" for what an electric field might be.

I have fond memories of 50 years ago standing on a breakwater protecting the boat basin of Santa Barbara's harbor in California. The water was clear with a tinge of blue green as it flowed past a sand bar at the tip of the breakwater. Picnicking on that sand bar and swimming in the protected water was a summer delight. The water was cool, but we were young and full of life. Looking seaward from the breakwater the waves appeared to arrive from a direction slightly to the north, reflect from the stone breakwater and depart at an angle slightly towards the south.

Although the wave peaks appeared to move, there was no net movement of the water, towards or away from me. Water did, however, move up and down due to stresses within the wave. These stresses, or up and down forces, did move towards and away from me, causing the apparent motion of the wave peaks and valleys. Stress forces migrate through the water because of a viscosity link between all small volumes of water.

Electric and magnetic forces in a traveling electromagnetic wave behave in a fashion similar the stress forces in a water wave. Up and down stresses in this case are due to oscillating electric and magnetic potentials. These potentials have what is called "influence at a distance." (ref 10) This influence serves the same as viscosity in a water wave. It transfers potential forces a small distance forward into space. This transfer occurs at the speed of light. We call a volume of space containing many potentials a "field."

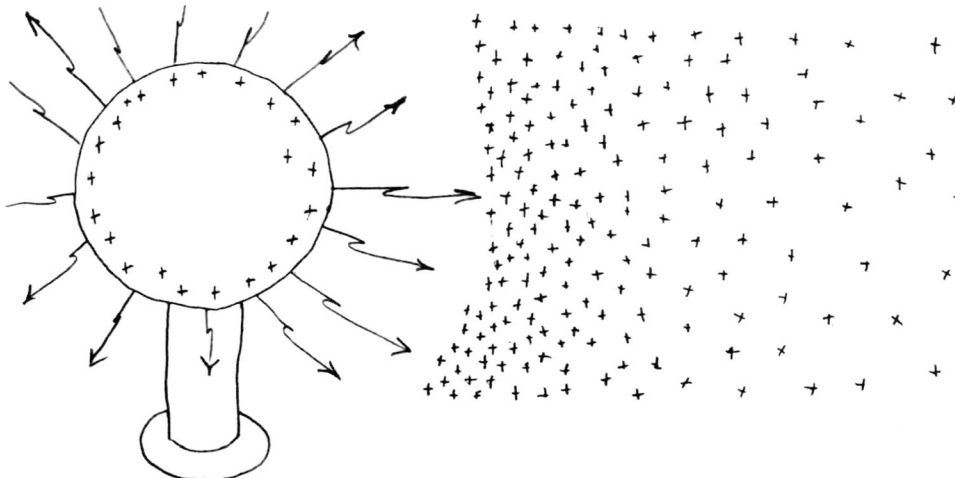

Fig. 2

In a static situation "influence at a distance" accounts for the fact that a large electric charge on a stationary object will be felt as an electric potential at a distance from that object. I will never forget the day a friend invited me to join him on a sail across Long Island Sound. It was a beautiful day and the trip started out well. There was a good breeze and the boat took the waves cleanly. We were almost across the Sound when a thunder cloud that had been threatening for some time caught up with us. I could feel electricity in the air. Hairs on my arm stood out. The rigging sizzled. St Elmo's Fire danced from the masthead and I feared being fried to a crisp by a lightning stroke. Now that was an electric field of the first order!

It will be shown later that once created, an independent electric or magnetic field will zip into space without further help. The question addressed here is, "Once created, do the companion electric and magnetic fields of electromagnetic energy also propagate into space without further action, or do they continue to "curl" around each other in accordance with Maxwell's symbiotic dance?"

Central to understanding light is the reader's willingness to accept as real the phenomena of "influence from a distance." Notice the change from "at" to "from." This is because, when thinking about a field at a point p, we are actually thinking about an influence from a source some distance away that has just arrived at point p. The idea of an influence, or force, from a distance should not be difficult to accept considering the fact that every human being has been under the influence of two all-encompassing fields since birth, earth's gravity and magnetic fields.

An influencing source can be close by. For example, if one holds a piece of iron next to a magnet the force generated by the magnetic's field coursing through the iron will be felt. Similarly, the force generated by two adjacent magnets can be hard to resist. Fields are real!

1.2 Early Discoveries

1.2.1 Magnetic Fields

Magnetism played a significant roll in the life of early times because of the magnetic compass, which was first used by Chinese sailors. Around 800 BC the Greeks named magnetic force after the rock magnetite, which had an attraction for iron. In the early 1600's W. Gilbert recognized that the earth is one large magnet. Several hundred years later a Dutch scientist, H. Oersted, found that an electric current in a loop of wire could cause a compass needle to turn. This was the beginning of an understanding of the intra-play between electricity and magnetic fields. By 1750 the north and south poles of a magnet and the nature of the force between them had been identified.

In the 1820's A. Ampere was among the first to quantify the behavior of magnetic fields and the ability of an electric current to generate one. He did the experiments that allowed him to write equation (1):

$$\int_S B \cdot dS = \mu_o I_s \quad \left(\frac{weber}{m^2} \right) \quad (1)$$

Where:

S	=	closed loop of wire
dS	=	a small integral of S
B	=	magnetic flux density inside S
Is	=	electric current through loop S
μ_o	=	a mathematical constant for dimensional correctness

Equation (1) is a mathematical equation that says a current, Is, through a closed loop of wire,

S, will generate a magnetic field, B, inside the loop. The summation, indicated by the integration sign, ∫, adds all of the field flux inside the loop S.

The sketch below shows how a magnetic field is defined when created by a fixed magnet. H is magnetic flux strength (ampere/meter), ϕ is the magnetic flux (weber) and B is the magnetic flux density (tesla). The subscript "t" indicates that the flux flows tangential to the wavefront, or perpendicular to the direction of propagation. In Figure 3 the magnitude of an isolated magnetic field is shown to disperse at a rate inversely proportional to the third power of the distance, r, from the source. This unexpected value is due to the fact that a magnetic source always has two poles. This third power will be confirmed by experiments described in Section 2.2.1.

About 1830, one of England's most well known scientists, M. Faraday, established his Law of Induction (ref 11):

$$emf = -\frac{d\phi}{dt} \quad (volt) \quad (2)$$

Where:

d	=	a small portion of
ϕ	=	magnetic flux (weber)
t	=	time (seconds)

Equation (2) says that moving lines of magnetic force, or magnetic flux, ϕ, will generate an electric force, emf, in a closed loop of wire. The term emf is the equivalent of the familiar "volt." The negative sign can not be explained, other than to say, "That's just the way this Law of Nature is."

The idea of equation (2) led Faraday to the invention of the electricity generator and soon after that to the invention of the electric motor, which inaugurated the industrial revolution in England.

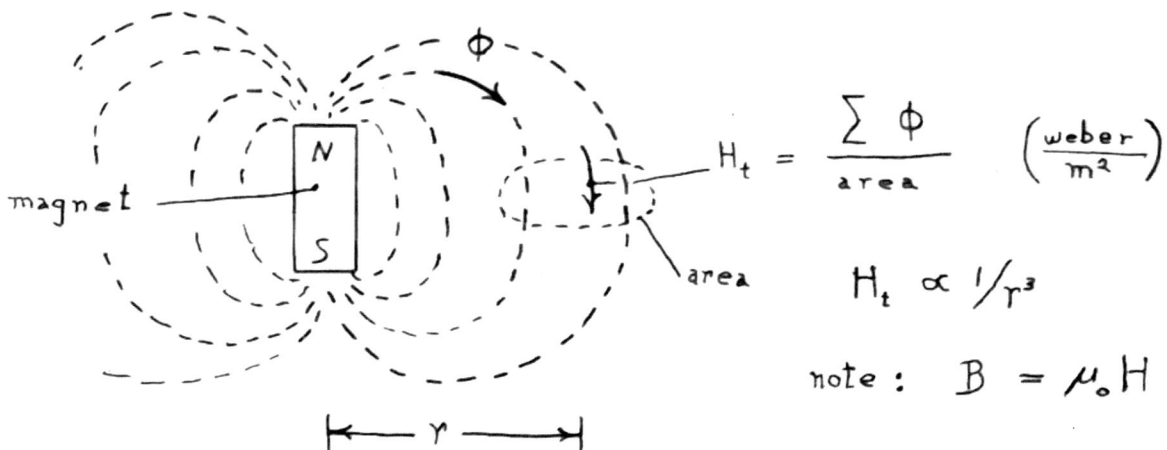

$$H_t = \frac{\sum \phi}{area} \quad \left(\frac{weber}{m^2}\right)$$

$$H_t \propto 1/r^3$$

note: $B = \mu_0 H$

Fig. 3

Figure 4 is from an early German school book and is probably a close copy of Faraday's electricity generator based upon equation (2), where a coil of wire passes through a magnetic field. M. Faraday also made a second key step towards understanding light with what he called "displacement field," which was defined by the following:

$$E = \mu_o \ \frac{d\,H}{d\,t} \ \left(\frac{volt}{m} \right) \qquad (3)$$

Equation (3) says that a magnetic field, H, that changes with time generates an electric field, E, slightly displaced from the H field. With equations (2) and (3) Faraday established his Field Theory, which says that field lines of force constitute "influence at a distance." (ref 12)

1.2.2 Electric Fields

The earliest use of electric effects, such as chasing floating feathers with an amber stick, were strictly for amusement. This was done by Greeks as early as 800 BC. This was also the case in 1600 when W. Gilbert introduced electrostatic trick parlor games. Benjamin Franklin was one of the first to examine electricity as a "natural philosophy" phenomena and in 1747 recognized that static and dynamic electric effects were due to elemental charges. A short time later an English scientist, C. Coulomb, established a quantitative description of the force between electric charges. He was the first to recognize the concept of an electric field, E, surrounding an electric charge, q, which is defined by equation (4):

Fig. 4

$$E \ \propto \ \frac{q}{r^2} \ \left(\frac{N}{C} \right) \qquad (4)$$

Where:

q	=	electric charge (coulomb)
N	=	force (Newton)
C	=	charge (coulomb)
$\propto$	=	means proportional to

The exponent of "r" indicated that field strength decays as the square of the distance, r, as one moves away from the source. Coulomb also described quantitatively the force, F, that exists between two electrostatic charges:

$$F \propto \frac{q_1 \cdot q_2}{r^2} \quad (N) \qquad (5)$$

Equations (4) and (5) embody the concept of "influence at a distance."

Figure 5 below shows how isolated E fields may be generated. It also demonstrates the difference between "radial" and "tangential" E fields. The flux of a radial field, Er, is parallel to the direction of propagation as the wavefront expands into space. The flux of a tangential field, Et, is parallel to the wavefront and perpendicular to the direction of propagation. The relationships between electric currents and magnetic fields, described by equations (2) and (3), were precursors and contributors to Maxwell's theoretical description of "light" fields.

1.3 Maxwell's Equations

By 1860 the intra-play between electric fields and magnetic fields was fairly well understood. A Scottish engineer, J. C. Maxwell, tied it all together. He combined the experimental work of Ampere and Faraday (who was Maxwell's close personal friend) with ideas from his own experiments, and arrived at the conclusion that "light is an electromagnetic disturbance." In 1873 he delivered a "Treatise on Electricity and Magnetism" which included equations (6) on the following page. These remain today the classical definition of electromagnetic radiation.

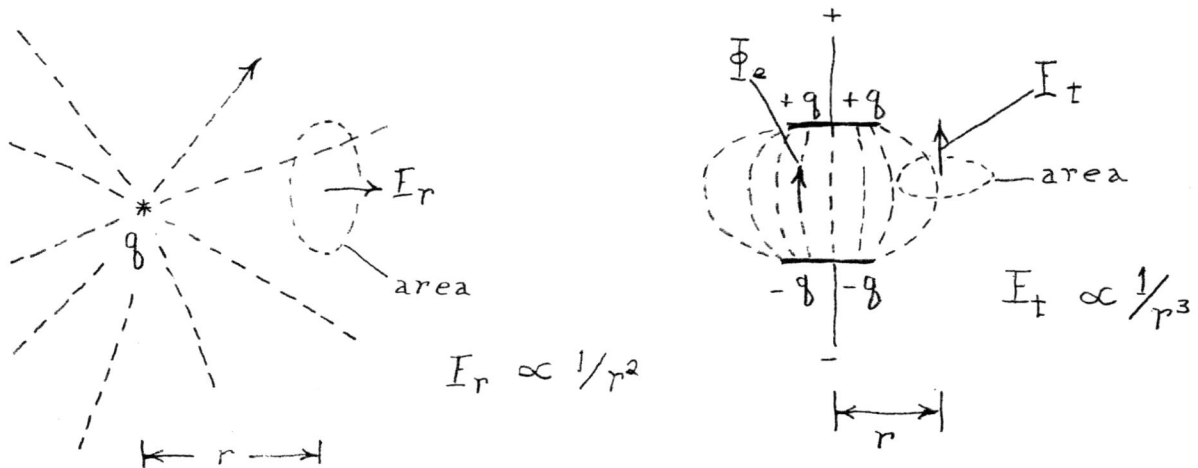

Fig. 5

$$\text{div } E \;=\; 0 \quad \text{or} \quad \frac{q_{in}}{\epsilon_o} \qquad (6a)$$

$$\text{div } H \;=\; 0 \qquad (6b)$$

$$\text{curl } E \;=\; -\frac{1}{v}\,\frac{\partial H}{\partial t} \qquad (6c)$$

$$\text{curl } H \;=\; \mu_o I + \frac{1}{v} \times \frac{\partial E}{\partial t} \qquad (6d)$$

$$v \;=\; \frac{1}{\sqrt{\epsilon_o \times \mu_o}} \qquad (6e)$$

Where:

v	=	velocity of propagation (m/sec)
I	=	electron currant (ampere)
q_{in}	=	internal charge
ϵ, μ	=	mathematical constants, values will become significant in Exp 7 and 12

The term div, divergence of electric flux from an enclosed volume, is always equal to 0 when there are no charges within the volume, or it becomes q_{in}/ϵ, when there are charges, according to Gauss's Law. The net div of magnetic flux always originates from a pair of magnetic sources, N and S.

Understanding the term "curl" is crucial to understanding Maxwell's description of light. The following arguments were taken from J. Newman's biography of Maxwell. (ref 13)

1- Curl in (6c) means an electric field curls around the fluctuating magnetic field that generated it.

2- Curl in (6d) is the counterpart for a magnetic field.

3 - The curl function is described as taking place in a tiny volume or point in space and an instant in time, therefore:

$$\frac{dI}{dt} \;\Rightarrow\; \frac{\partial E}{\partial t} \quad \text{and}$$

$$\frac{dH}{dt} \;\Rightarrow\; \frac{\partial H}{\partial t} \qquad (7)$$

4 - Equations (6c) and (6d) describe the self-sustaining nature of light. A fluctuating magnetic field generates an electric field, which in turn generates a magnetic field. E and H fields

form at right angles to each other, and both are at right angle to the direction of propagation. Thus light propagation in free space is not only self-sustaining but also lossless.

Equation (6e) led Maxwell to realize that his "v" had the same value as the current value for the velocity of light. Because of this he identified light as an electromagnetic phenomena.

The symbol ⇒ has been introduced to mean "causes" or "results in," thus making the flow of thought to coincide with the flow of action being described.

Maxwell's third equation is his interpretation of Faraday's Law of Induction (2) which could be written as:

$$\frac{\partial H}{\partial t} \Rightarrow -\oint_s E \cdot \partial s$$

$$\text{or} \quad curl -E \qquad (8)$$

His forth equation is a combination of Ampere's equation (1) and Maxwell's "displacement current" and could be written as:

$$I + \frac{\partial E}{\partial t} \Rightarrow \int_s H \cdot \partial s$$

$$\text{or} \quad curl\ H \qquad (9)$$

When originally presented, Maxwell's equations were not well understood and were rejected by his contemporaries. It was not until several years after his death that his ideas were accepted. At that time H. Hertz interpreted his work and demonstrated with radio waves the existence of electromagnetic radiation.

The self-sustaining nature of light propagating by successive electric field curl followed by magnetic field curl is still taught in today's physics text books, but experiments described in subsequent sections of this book will show that curl occurs only at the generation of light, and not during it's propagation.

"The grand overriding law of the parsimony of Nature: every action within a system is executed with the least expenditure of energy."

- Maxwell, 1873 (ref 14)

2. EMISSION

Although the transition from vibrating electrons to electric and magnetic fields of propagating light waves, as described by Maxwell's equations (6), has not been treated in textbooks devoted to classical physics, there are clear descriptions of how electrons vibrating at radio frequencies generate E and H fields. Section 2.1.1 contains concepts and further ideas concerning this radio emission. Section 2.1.2 then presents ideas about the same phenomena for visible-light wavelengths.

Further curiosity concerning this transition from vibrating electrons to propagating E and H fields led us to search experimentally for a missing link. Although no correlation was found between the experimental results and known characteristics of propagating radio waves, it became quite clear that in close proximity of a radiating antenna there exist multiple wave forms. In some cases the confusion brought on by these wave forms boarders on chaos. Radio engineers refer to this region as "near field." A few of these "near field" experiments using audio and radio frequencies are described in Section 2.2.

As one moves away from the radiating antenna, isolated "coherent" waveforms are found. This region is known as the "far field." Far field electromagnetic coherence is described in Section 2.3.

2.1 Concepts

2.1.1 Radio waves

Compared to light waves, radio waves generally originate from a single source and are relatively coherent. Thus field structure in the vicinity of a radio antenna can be well defined. Figures 6 through 8 contain the wave forms for the simplest case of highly coherent radiation from a single pole, where the pole height is much less than one wavelength. They cover the transition from moving electrons to propagating isolated fields.

Figure 6 shows the meaning of Maxwell's term "curl" by means of three steps.

In step 1, a current flows into the pole. This current causes an H field to curl around the pole. Note that this H field flux direction, i.e. vector, is at a right angle to the I vector.

In step 2, the current flow has stopped, which causes the H field to collapse. This in turn causes an E field to curl around the pole. This E field vector is at a right angle to the H field. Both E and H vectors are at right angles to a radial vector originating from the pole.

In step 3, we see a horizontal slice through the radiated field. Several successive curling circles are shown frozen in time. As in steps 1 and 2, field flux flows perpendicular to the radial vector, but field influence in the form of a curl field will be felt at a small distance along a radial vector. There are many sequential curl H, curl E, curl H, and curl E circles surrounding the pole, each a small radial distance beyond the last. Therefore we may call a radial vector a propagation vector. The distance between successive H field curls is the wavelength of the propagating energy. And further, because the field curls are at right angles to each other, there is no energy lost in the propagation of

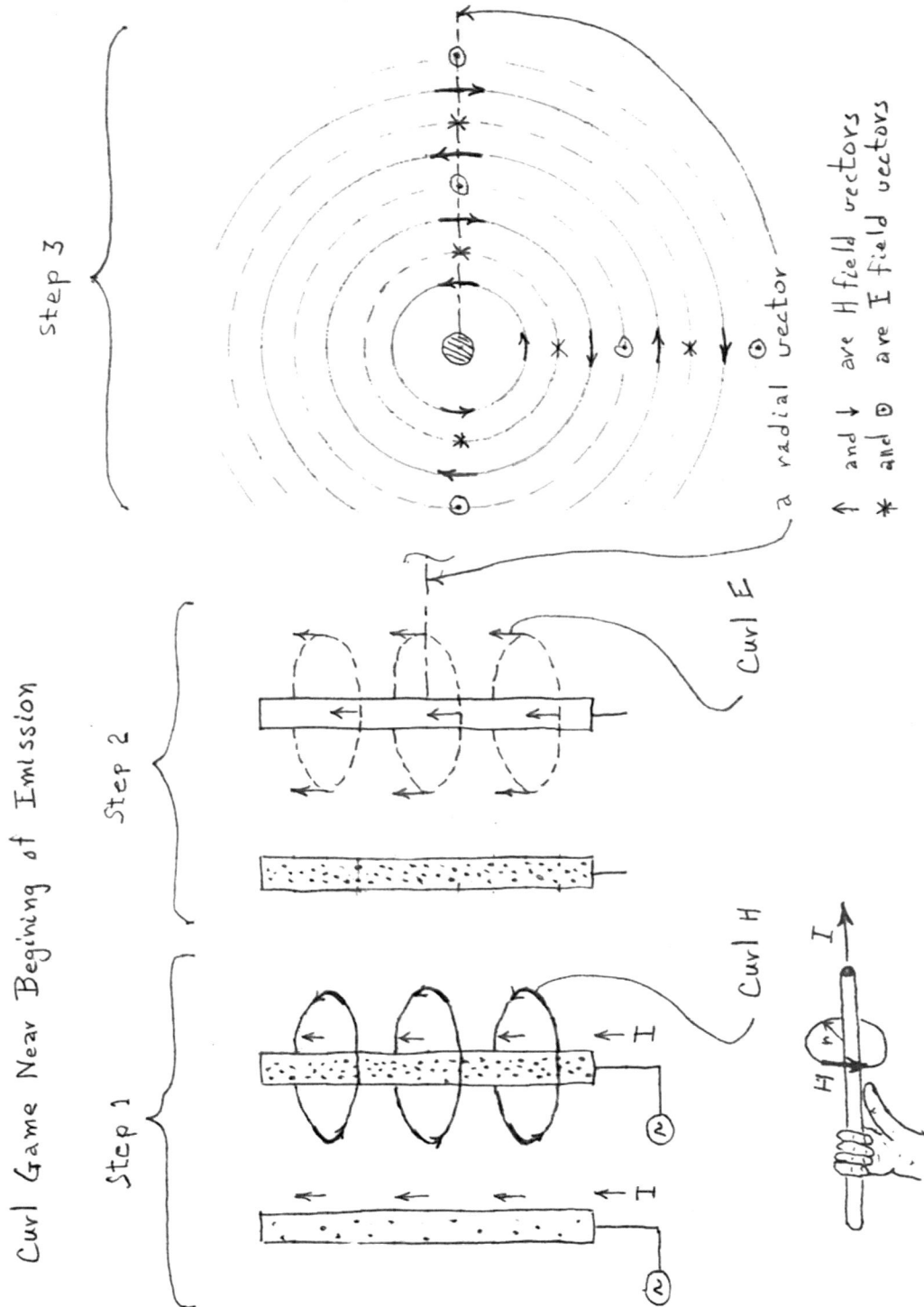

Fig. 6

Curl Game Near Begining of Imission

step 3

Step 2

Step 1

Curl E

Curl H

a radial vector

↑ and ↓ are H field vectors
✳ and ⊕ are E field vectors

↑ a radial vector
✳

I

H

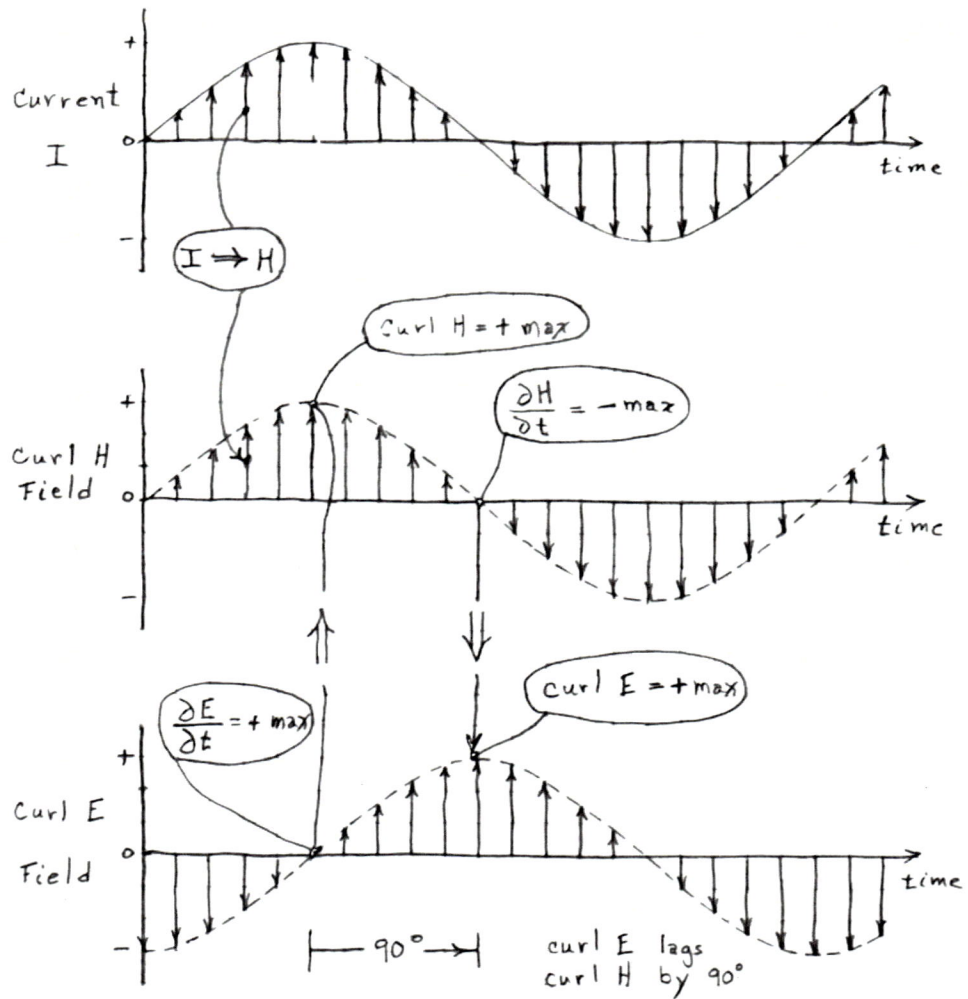

Fig. 7

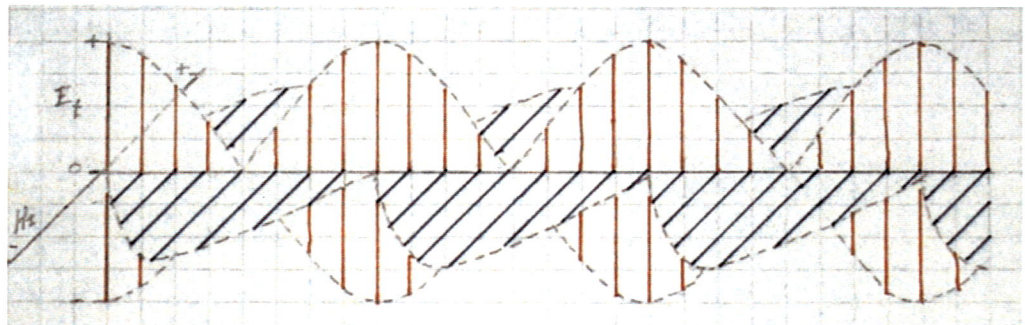

Fig. 8

light in a vacuum. There is however a $1/distance^2$ dispersion of energy as it propagates through space.

Figure 7 is a slice in frozen time of the dipole current and the H and E curl fields along a propagation vector as in Figure 6. Referring back to Maxwell's equations (6c) and (6d), it can be seen that he anticipated that a curl E would be generated by a negative change of H with time, $-\partial H/\partial t$, and a curl H would be generated by a positive $\partial E/\partial t$. Note that in Figure 7 a negative $\partial H/\partial t$ causes a positive curl E per equation (6c), and a positive $\partial E/\partial t$ causes a positive curl H per equation (6d). Figure 7 shows why, in principle, these sequences cause a propagating E field to lag behind it's companion H field by 90 degrees.

In addition to the time-phase difference between E and H fields there is also a polarization-phase difference of 90 degrees between the two fields. Figure 8 contains a composite display of E and H fields combined with both the time-phase and polarization-phase differences.

Maxwell visualized that the curl action described here, when applied to the generation of light, occurred in minute space and time. It is indeed minute because we believe that only energized electrons are capable of generating light.

2.1.2 Visible Waves and Photons

2.1.2.1 Planck's Constraint

Perhaps the most significant of Nature's laws concerning light is that conceived by Max Planck in 1900. At the time he was searching for an equation that would account for the spectral characteristics of infrared radiation. He succeeded in his quest by adopting a computational convenience from Boltzman's statistical analysis. This led to a surprising conclusion that "atomic resonators emit only discrete elements of energy." Planck's simple equation for the emission of electromagnetic energy is:

$$\xi e = m\,h\,v \quad \text{(joule)} \qquad (10)$$

Where:

ξe	=	emitted energy (joule)
h	=	Planck's constant
v	=	quanta frequency (1/sec)
m	=	number of discrete elements

Planck's discrete element of energy, when $m = 1.0$, was subsequently referred to by Einstein as a "quanta." It is said that this marked the beginning of quantum physics. Equation (11) below says that the smallest quantity of light, ξ, is : (ref 16)

$$\xi = h\,v \quad \text{(joule)} \qquad (11)$$

"If I plunge the end of a red hot bar of iron into a vessel of water, the fluid rises, ... Why may we not, for a moment, compare the rays of the sun, darted directly upon the surface of the water, to many bars of red hot iron; each bar, indeed, infinitely small, but not the less powerful? In this case wherever a ray of fire darts, the water will be driven on all sides ..."

- Oliver Goldsmith, 1779 (ref 15)

The value of ξ in equation (11) is actually a constraint limiting the minimum energy of a quanta at emission. The value of v is equal to the inverse of the quanta's frequency. Subsequent sections will show that an electron is always involved in both the emission and detection of light. Because of this the author believes that Planck's constraint is an intrinsic characteristic of the electron. This is not a new idea. It was also suggested by Robert Millikan almost 100 years ago:

> *"Plank's h seems to be always tied up in some way with the emission and absorption of energy by the electron. h may therefore be considered as one of the properties of the electron."*
>
> *- R. Millikan, 1917* (ref 17)

2.1.2.2 Bohr's Model

N. Bohr used Planck's constant to help explain the structure of the hydrogen atom and why, when excited, it radiates light with a characteristic wavelength. Bohr's model of the atom is a simple combination of old and new Physics. It combines Coulomb's electrostatic law and Newton's law of motion and the concentration of energy, with Planck's constraint in a theory that predicts the wavelength of hydrogen radiation to a remarkable precision.

He assumed that a hydrogen atom consisted of a large positively charged proton surrounded by electrons in orbit around it, much like our solar system with planets orbiting the sun. For a stable electron orbit, the forces on the electrons must be in perfect balance. By virtue of its being in orbit around the proton, the electron has two forms of external energy. It will have kinetic energy, K (r), due to its mass and motion in orbit. And it will have potential energy, U (r), due to the electrostatic charges on it and the proton. The electron's total energy depending on r, orbit radius, can then be expressed by:

$$E\,(r)\ =\ K\,(r) + U\,(r) \qquad (12)$$

As it stands, this expression allows r to take on any value.

This is the point at which Bohr's instinct took over. Knowing that Planck's constant defined the smallest unit of electromagnetic energy allowed, he guessed that Nature had imposed a similar limit on the smallest increment of energy allowed the orbiting electron in the hydrogen atom. He applied this limit as a constraint on electron energy expressed by:

$$2\,\pi\,r\,\cdot\,m\,v^2\ =\ n\,\cdot\,h\upsilon \qquad (13)$$

Where:

m = electron mass (m)
v = electron velocity (m/s)
r = electron orbit radius (m)
n = integer number

Why Planck's constraint is valid is not known. However it is known that when Bohr combined equation (12) with equation (13) the result was an equation that led to a correct calculation of observed hydrogen radiation spectral line:

Calculated
Spectral Wavelength = 1214 angstrom
Measured
Spectral Wavelength = 1216 angstrom

Although Bohr's model of the atom permitted calculation of the wavelength of light emitted by a single hydrogen atom, exact calculations for multiple electron atoms became exceedingly difficult. When excited, larger atoms emit energy at a number of discrete wavelengths. Figure 9 is from an experiment conducted by H. Kirkpatrick and the author while at college. It shows multiple lines emitted by several atoms, including mercury. Note that it also shows that a GE fluorescent lamp contains mercury. Perhaps this is why we were told not to breath the fumes from a broken fluorescent lamp.

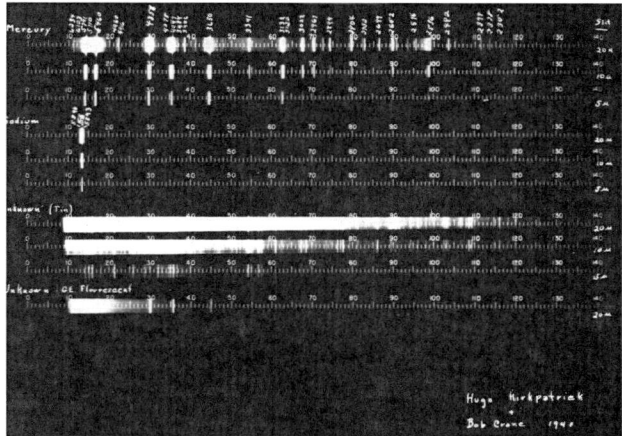

Fig. 9

2.1.2.3 Birth of a Photon

Remarkable as it is, Bohr's model of the atom does not account for the fact that an emitted photon is actually a small packet of waves. Some say that Planck's equation (11) says it all. But we believe that

Maxwell's parsimony of Nature demands a gentle electron jump, i.e., a damped oscillation between orbits or energy levels, as pictured in Figure 10 below.

Figure 10 is the Bohr model of a neon atom modified to show an electron oscillating between an outer shell, or excited state, and a stable state. And thus, "Anon the atom lets flee a photon."(ref 18)

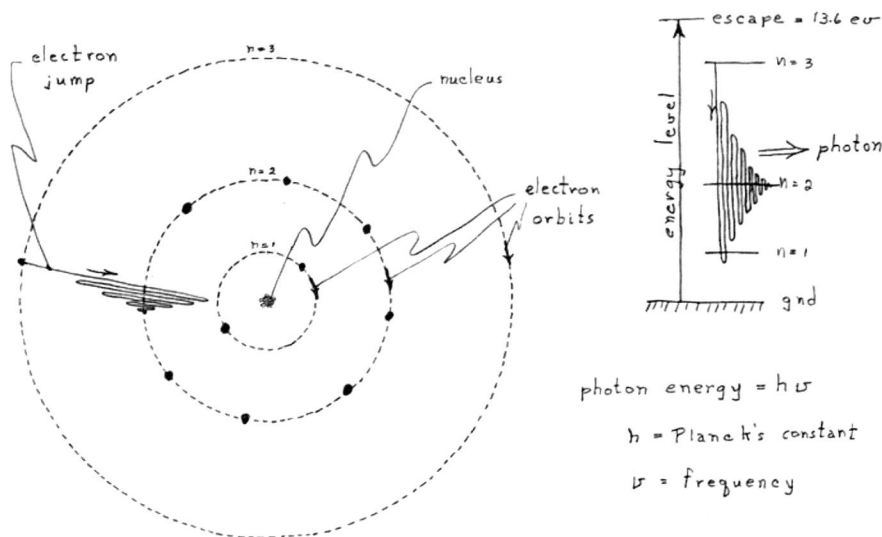

Fig. 10

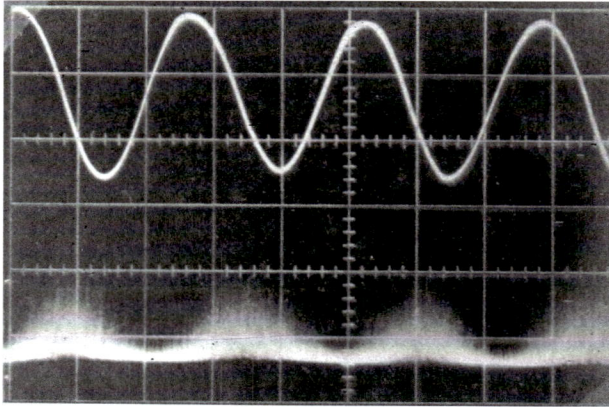

Fig. 11

The wave theory of light versus the photon theory of light engendered great controversy during the 1800's. It is now well understood that light photons possess wave characteristics and that waves possess impulse characteristics. In other words, light waves and photons are the same thing. If there remains anyone who doubts this dichotomy, please look carefully at Figure 11. This is a photograph of a dual-trace oscilloscope display taken during a two-beam interferometer experiment. In this experiment the optical frequency of one beam was changed very slightly. The interferometer output beam was therefore modulated at the difference frequency.

The output beam was detected by a photo multiplier tube. The electric signal was fed directly to one channel of the oscilloscope, bottom trace. The same signal was fed through a low-pass filter and then to the oscilloscope, top trace. Note that the bottom trace shows the impulse nature of the light due discrete quanta, and the top trace shows the sinusoidal beat note due

to the wave nature of the light. Thus Figure 11 demonstrates the compatibility of Planck's "discrete elements of energy," Section 2.1.2.1, and the interference of two beams of light, Section 2.3.2.

If the foregoing ideas are close to reality then a quanta is a decaying set of alternating magnetic and electric fields with duration dependent upon local conditions and frequency dependent upon specific atomic energy levels. Such a model is consistent with both the quantum and wave theories of light, and furthermore might look something like the photograph below which I took of a passing quanta while I was flying very high in a high-speed aircraft (!).

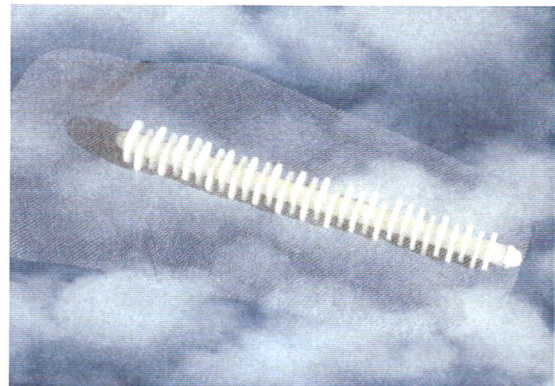

Fig. 12

2.2 Near Field Experiments

This section describes experiments performed in search of an empirical understanding of the missing link between vibrating electrons and propagating E and H fields. The specific objectives were to generate individual E and H fields possessing individual characteristics similar to those of propagating light. The following characteristics were selected:

1. Propagation dispersion loss

For simplicity the following propagation loss law was selected:

$$E \text{ or } H \propto r^n \qquad (14)$$

Where:
E = electric field intensity
H = magnetic flux intensity
r = range or distance from the source

When the logarithm of intensity is graphed versus the logarithm of range, the slope of the curve gives the value of "n." Electromagnetic energy propagates with a value of n = -2. Hence the prime objective of near-field experiments was to find those fields in which n = -2.

2. Evidence of "curl, curl, curl"

Generation in space of an H field by an alternating E field, or visa versa, would be evidence of Maxwell's "curl."

3. Susceptible to "draw down," sometimes called "suck in"

For a long time the author has wondered how a very thin antenna wire or a very tiny electron could have sufficient cross-section area to absorb a detectable amount of energy from propagating waves that clearly extend over an area thousands of times their size. It appears as though there is some kind of hokey poky going on when a small wire or iron core "draws down" energy from relatively large fields.

The draw-down effect can be understood by recognizing that a radio antenna, for example, removes electric energy from a volume containing an electromagnetic field. This, in turn, distorts the electric field. Thus, when the E field bends, the electric flux vector bends to bring more energy to the antenna. Draw down of an electric field should come as no surprise considering the success of Benjamin Franklin's lightning rods in 1750. Draw down will be discussed in more detail in Section 4.2.

2.2.1 Magnetic Field Experiments

Experiment #1a: H Field Propagation Coefficient

The first of three experiments, as shown in Figure 13 on the following page, measured the force of a magnetic field as a function of r, range. The source of the field was a solid cobalt magnet. Field strength was measured by comparing it to the strength of the earth's magnetic field, approximately 1/2 gauss in Connecticut. Relative strength of the two fields was indicated by the angle of a compass

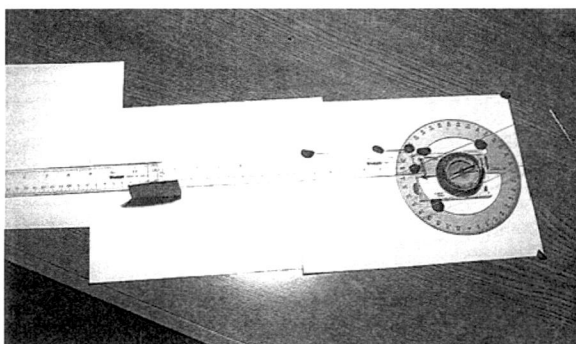

Fig. 13

needle subject to both fields at right angles to each other. A sketch showing the arrangement geometry is shown in Figure 14. Results of the experiment are graphed with log-log scales. The slope of this graph is $n = -2.95$, which according to equation (14) means that the field strength is very close to inversely proportional to r^3. It was anticipated that a value of $n = -2$ would be found.

It will later be found that all fields generated by dual sources, such as N and S poles of a magnet, will have a propagation constant of -3.

Experiment #1b: H Field Propagation Coefficient

A second magnetic field experiment, Figures 15 and 16, also measured the propagation coefficient, i.e. the value of n, for a magnetic field as a function of distance, r. In this case an alternating field was generated by a rotating cobalt magnet. The receiver was a multi-turn coil which generated an alternating voltage in accordance with equation (3). This voltage was displayed on an oscilloscope. Results are graphed in Figure 17 on the

following page, with log scales and shows a value for n of -2.86.

Use of a rotating magnet as a source of magnetic field had several advantages. First and foremost it generated a pure modulated field. Rotation was caused by a falling lead weight. Thus there were no spurious fields, such as plagued many subsequent experiments.

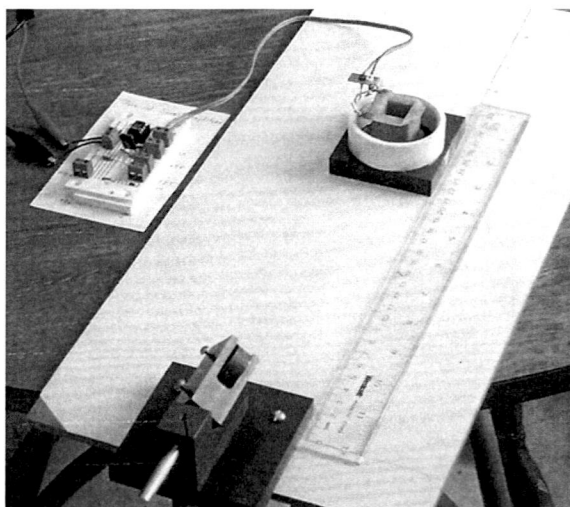

Fig. 15

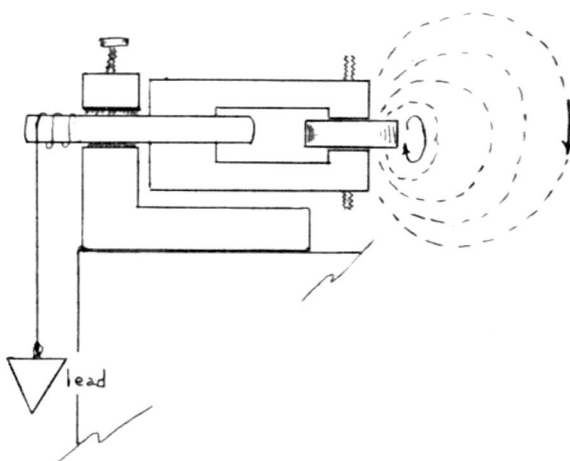

Figure 16

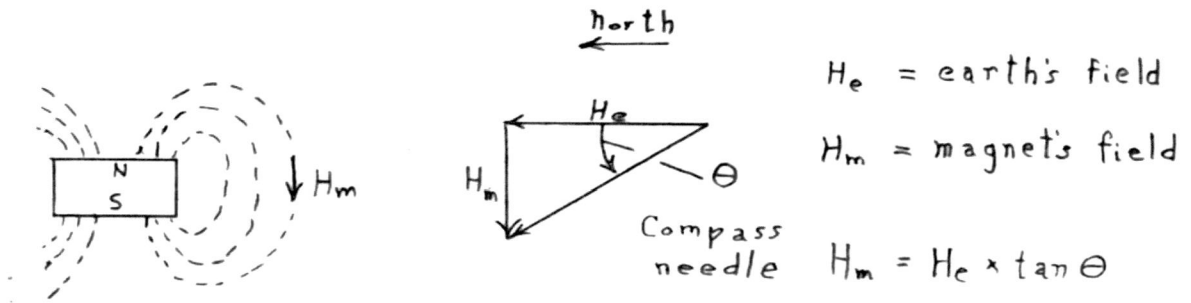

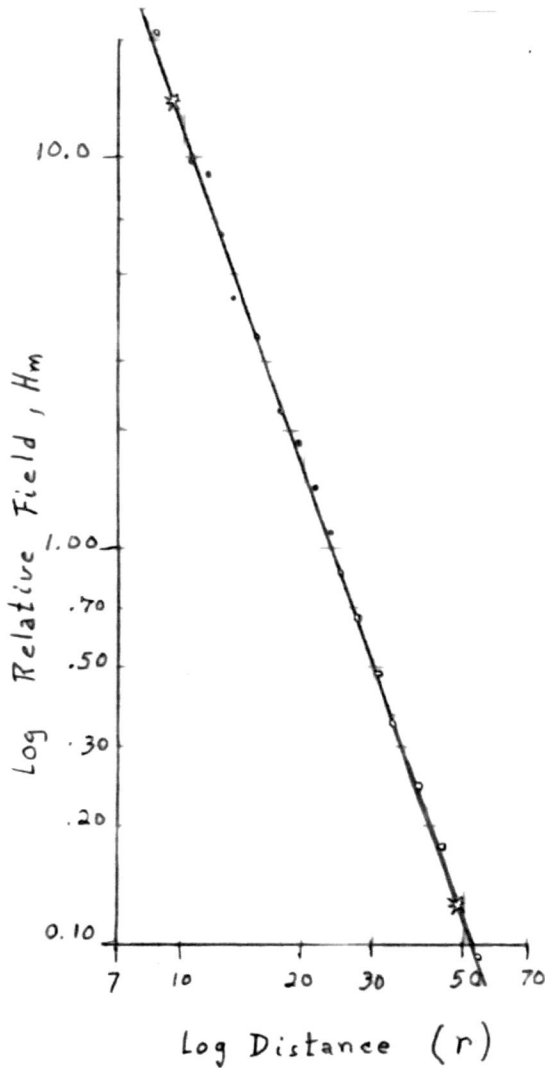

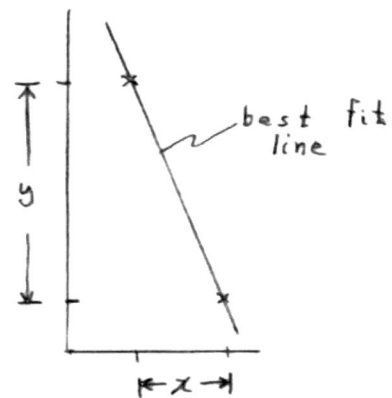

north

H_e = earth's field

H_m = magnet's field

Compass needle $H_m = H_e \times \tan \theta$

slope = y/x

slope = -2.95

$\therefore$ $H_m \propto 1/r^{2.95}$

Fig. 14

Unfortunately this model generated only very low frequency modulation, which considerably reduced it's effectiveness.

Experiment #1c: H Field Propagation Coefficient

A third set of magnetic field tests, Figure 18, was conducted at a higher frequency in which fields were created by a current-driven multi-turn coil in accordance with equation (3), in hopes of creating a stronger field.

These tests were performed over ranges up to six feet and at frequencies up to 1500 cps. Typical values for n were -2.8. Clearly such a magnetic field is not similar to that associated with a propagating radio or light wave.

Figure 19 is included to show how simple equations can be used to lend credibility to empirical results from an experiment. The schematic diagram in the top portion of the figure describes Experiment 1c. Engineering data such as voltage levels, coil turns, tuned circuit Q and op am voltage gain would be needed to establish numerical credibility. But that would be an engineering class, not 1st or 2nd year physics. However it is important to have and understand physics laws and their relationships showing what is happening in an experiment. These are given in the lower half of Figure 19. It is nice to find that laws of physics derived over 100 years ago are still effective today.

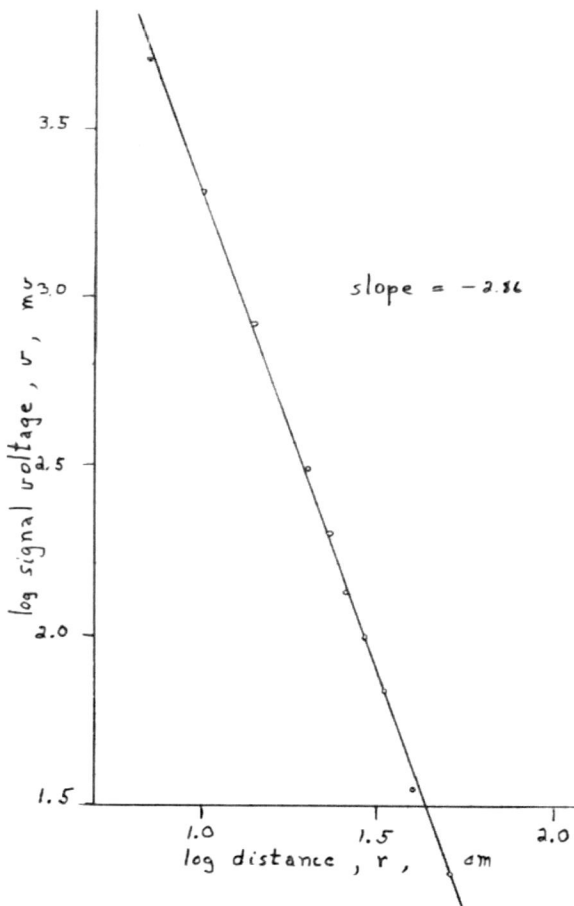

Figure 17

Figure 18

Objective: Determine value n in:

$$\Phi_m \propto r^n$$

where: Φ_m = magnetic flux (weber)
r = distance from source (cm)

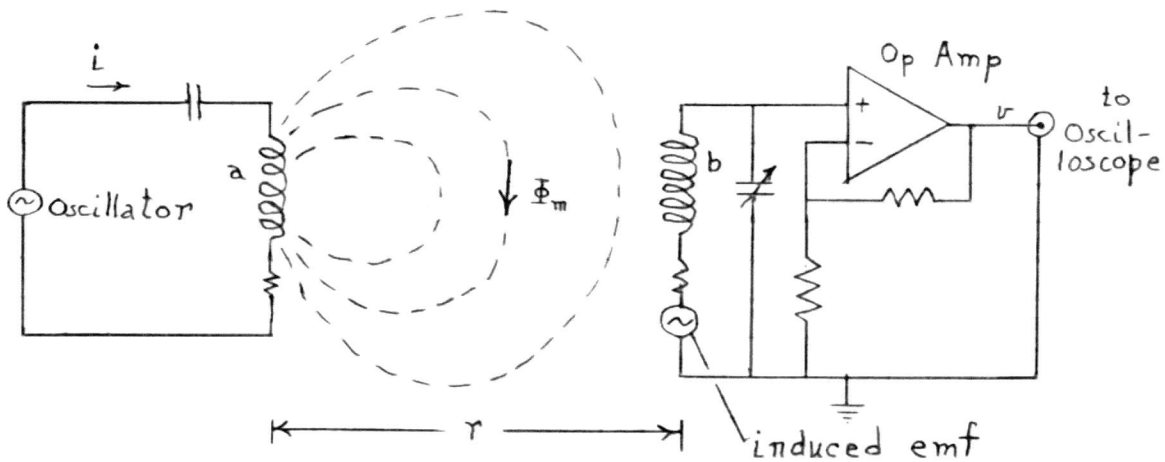

Ampere's Law for coil a

$$\Phi_m = \oint_S B \cdot dS \Rightarrow \frac{N}{\ell} \mu_0 \, i \quad (1)$$

where: N = # turns
i = coil current
μ_0 = a constant
ℓ = coil length

Faraday's Law for coil b

$$emf = -N \frac{d\Phi_m}{dt} \quad volts \quad (2)$$

where: N = # turns

Figure 19

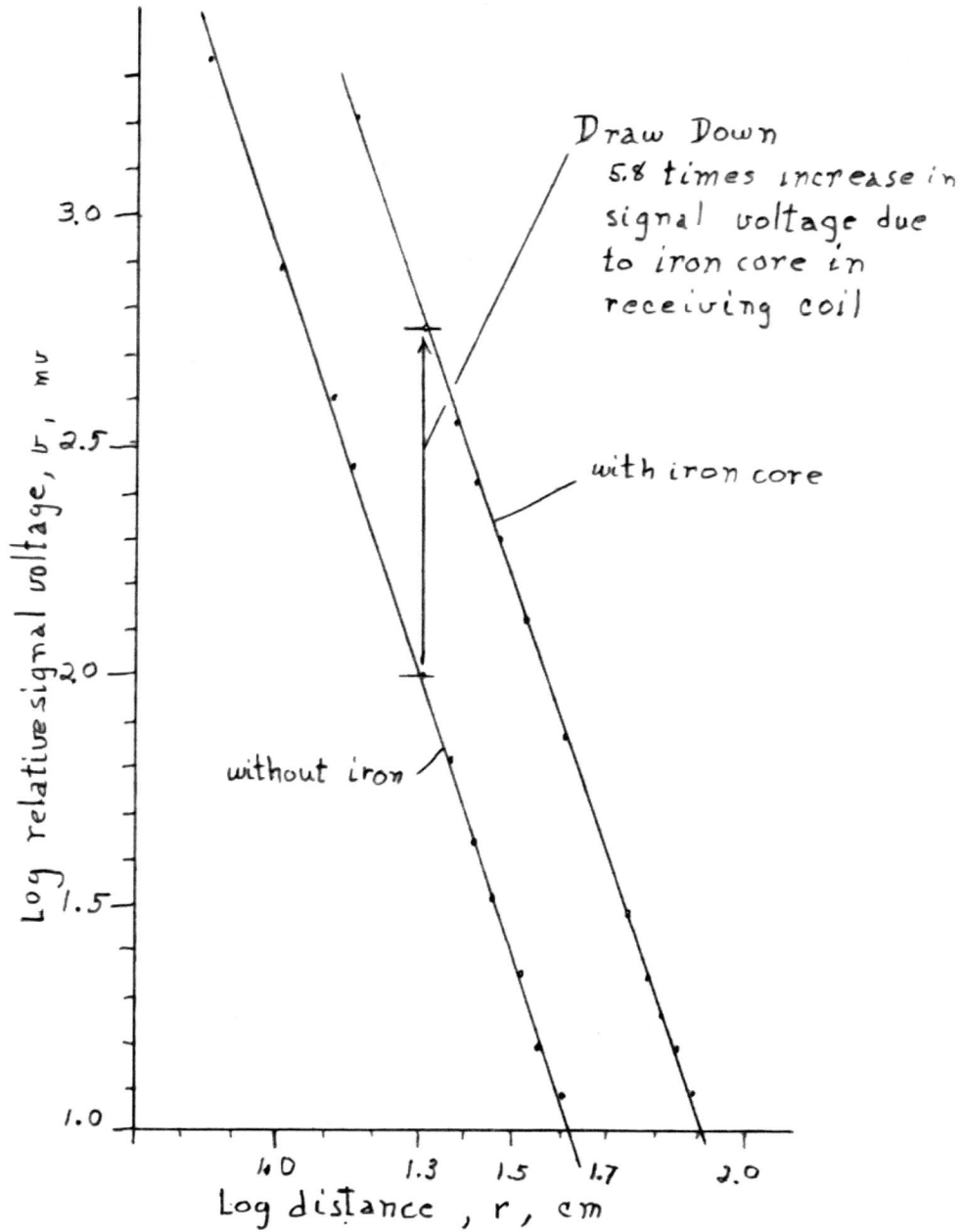

Figure 20

Experiment #2: H Field Draw Down

Experiment 2 was designed to show magnetic field draw down. Results graphed in Figure 20, to the left, are from two measurements of n for magnetic fields. Both used a transformer coil receiver, one with an air core and one with an iron core. The iron core drew in 5.8 times as much magnetic flux from the field as the air core. This caused the field to be warped or bent away from line-of-sight propagation. Section 4.2 will further describe field draw down.

2.2.2 Electric Field Experiments

Experiment #3: E Field Propagation Coefficient

The exclusive use up until now of the symbol E to represent the electric field is a little misleading. This is because essentially all light detectors* respond not to the strength of an E field (volt/meter), but rather to the E field potential, V (volt). E field potential at a point p in a field, is defined by the amount of energy required to move an electron from very far away to point p. Mathematically this is described by the following:

$$V = k q / r \quad \text{(volt)} \qquad (15)$$

Where:
k = a constant for dimensional correctness
q = the charge (coulomb)

r = distance from the charge to point p (meter)
Field potential is important here because the strength of V indicates the field's potential ability to generate a voltage in a wire. All Et field experiments involved measurement of V. Note that the exponent of r in equation (15) is n = -1, which corresponds to n = -2 for Et. Note also that the time phase of V is always the same as that for the corresponding E field.

Several experiments were performed in search of a propagation coefficient for V of n = -1. A variety of emitter and receiver arrangements were used. Photos on the following page show typical radiators and receivers and the calibrated track used. In all cases the receiver was fixed and the emitter was moved to calibrated distances. It was found impractical to generate a simple Et field, and only Er fields were used. Graphs of four typical but inconclusive experiments are given in Figure 23 on the following page. Slopes ranged from -1.3 to -1.6 with an average of -1.5. It is clear that the graphs do not exhibit propagation characteristics similar to those of electromagnetic radiation.

> *"Science is not one success after another. It's mostly one success in a desert of failure."*
>
> - *Folkman* (ref 19)

---

*not including thermal detectors

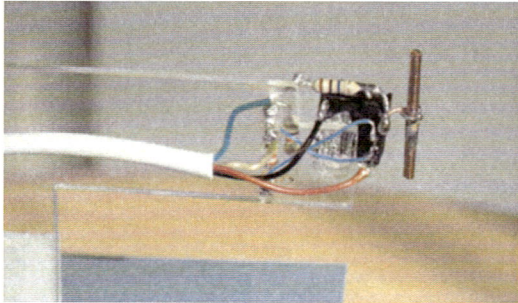

Fig. 21a.

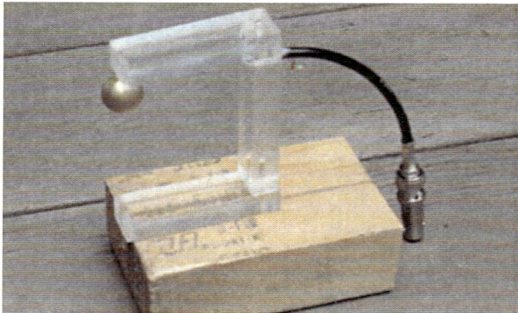

Fig. 21b.

Fig. 21c.

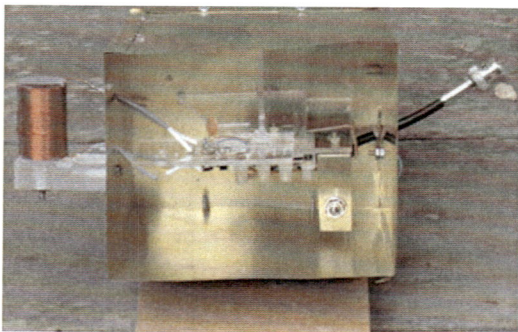

Fig. 21d.

Figure 21a. Receiver with a short rod antenna for V reception

Figure 21b. Single-ball radiator for Er emission

Figure 21c. Two-disk antenna for Et reception

Figure 21d. Bifilar coil antenna with shielded pre-amp for Ht reception with no response to spurious Er fields

Figure 22. Receiver with short rod antenna and a rod type emitting antenna mounted on a section of the range track. Just showing are the holes at calibrated ranges for the location pins, such as that showing in Figure 21c, protruding from the bottom of each receiver and emitter.

Although the search for a propagation coefficient of n = - 1 was disappointing there were two unforeseen and interesting results. The first of these was a clear demonstration that autonomous Er fields can be generated and sustained in propagation without Maxwell's "curl." They are self sustaining and expand forever in the most parsimonious way. The second was that the same can be said for Ht fields.

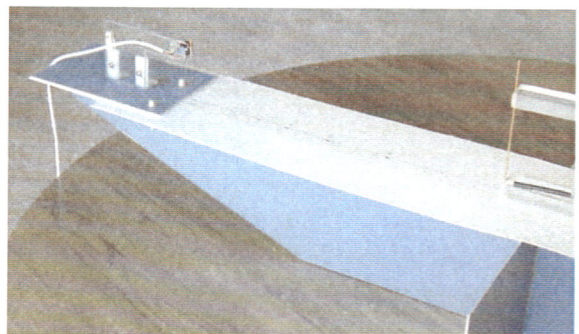

Figure 22

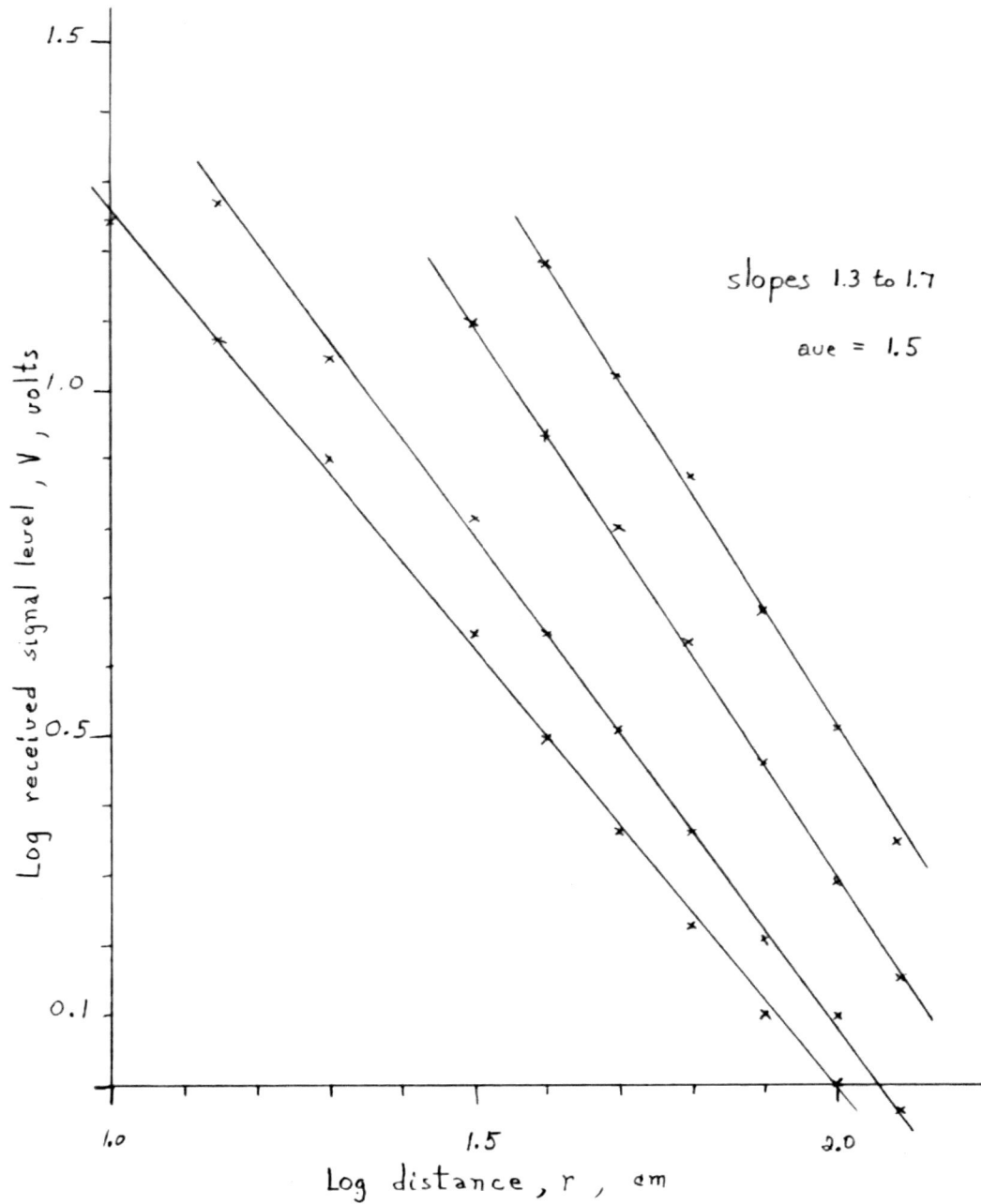

Figure 23

2.3 Structure of Light

2.3.1 Definitions

Quanta: The discrete element of energy defined by Planck's constraint. A quanta consists of a train of cyclic E and H fields. The length of this train determines the spectral bandwidth of the quanta; a long train results in a narrow bandwidth, a short train results in a wide bandwidth.

Field: The volume throughout which a large quantity of quanta have the same optical frequency.

Coherence: The volume of a field within which a large quantity of quanta have both the same frequency and essentially the same phase. Coherence is required for recognizable interference to take place.

2.3.2 Interference

When two beams of light occupy the same space, E field voltages add or subtract. This is called interference. Normally the net voltage changes so rapidly they are not recognizable. When the voltage sums or differences can be recognized the beams are said to be coherent with each other. A recognizable pattern made by the summation is called a fringe pattern. Bright areas of a pattern are due to "constructive" interference and dark areas are due to "destructive" interference.

Interference is a lossless event. When constructive interference takes place a light detector measures the sum of the two E field intensities. When destructive interference occurs a light detector measures the difference between the two E field intensities,

Fig. 24

one positive and one negative. This difference can be as low as zero, but neither field is destroyed, there is no loss of energy. The fields can not be measured by an E field detector, but both beams are still there.

Six examples of different kinds of interference patterns may be found on the front cover. Most notable is the large circular pattern at the center, called a zone plate. This pattern was made by a two-beam interferometer in which one beam reflected from a good flat and one reflected from a slightly spherical flat. This pattern is identical to that made by Isaac Newton using a flat and a lens surface. In spite of the obvious fact, Newton refused to accept that light can behave as a set of waves.

2.3.3 Coherence

If one thinks for a moment about the vast number of autonomous photons, or quanta, emitted from an ordinary light bulb, or from the sun, a question comes to mind. What influences such a large group of quanta, which have been issued at random time and a random spatial distribution, to organize themselves into pristine three-dimensional waves such that interference patterns can be observed?

It is somewhat similar to waves on the surface of a large expanse of water. They start as many wind driven ripples. With wind continuous over time they eventually form into large majestic waves. Waves such as those I observed while standing on the breakwater protecting the Santa Barbara harbor. In this case the forming influences are

surface tension and water viscosity combined with Maxwell's "grand overriding law of the parsimony of Nature."

What are the influences that convert a chaotic cloud of photons surrounding a light source into symmetrical waves of light? I suspect that again it is Nature's parsimony law augmented by an electric field's "influence at a distance," which has been described earlier. The coherence of light defines the physical extent of this influence.

Influence at a distance is more prevalent in Nature than we realize. For example, on the shore of North Carolina, close to the city of Wilmington, there is a beautiful sandy beach where Nature's mysteries abound. There is one mystery in particular that intrigues the author. It involves what he calls sand dabs, a small crustacean about the size of the last joint of the small finger. Sand dabs live in large colonies just below the sand surface on that portion of the beach washed by the back flow of dying waves. The water rushes up and then gently flows back. Sand dabs extend fan-like feelers to extract minute particles of food from the water. This goes on all day, and all night too. The mystery occurs as the tide recedes. Soon the water above the sand dabs is too shallow. Suddenly the sand erupts, and hundreds of sand dabs in unison pop out of the sand, scurry towards deeper water and just as quickly bury themselves in the sand again. Who blows the whistle? What influences the entire colony to simultaneously jump to a frenzied dash to deeper water?

The coherence of light is defined here as the distance along (axial) and across (lateral) a light beam over which samples may be extracted from the beam and subsequently combined to produce detectable interference fringes with a contrast of $\geq$ 50%. Numerical values of coherence place a limit on the effective optical aperture and the optical path difference of an interferometer, which extracts two samples from the optical beam and superimposes the samples to produce interference fringes. The author has used the following definitions of coherence for a number of years, working in the field of optical interferometry.

$$ \ell_z = \frac{1}{4} \; \frac{\lambda}{\Delta \lambda} \times \lambda \quad (16) $$

$$ \ell_{x,y} = \frac{1}{2} \; \frac{D}{S} \times \lambda \quad (17) $$

Where:

lz	=	axial coherence	same
lx,y	=	lateral coherence	dimensions
λ	=	wavelength	
$\Delta \lambda$	=	spectral bandwidth	
S	=	source size	same
D	=	distance to source	dimensions

Note that equation (16) indicates that a narrow band of light will have a long axial coherence. In the extreme, a laser can be designed to generate light at essentially a single wavelength, which results in very long axial coherence.

Experiment #4: Coherence Measurement

Figure 25 shows results from Experiment 4. It contains three interferograms made from light with three different coherence lengths. Trace a, Figure 25a, was obtained with simulated sunlight which had a very short axial coherence length. Trace b, Figure 25b, was made using light from a tungsten arc lamp. And Trace c, Figure 25c, was made using light from a HeNe laser, which has a very long axial coherence length.

The sketch on page 30, Figure 26, shows the optical arrangement used to obtain this data. Fringes were generated by the scanning mirror which changed the optical path difference between the two reflecting mirrors. As this difference changed the interference between the two beams cycles between constructive and destructive. Peak intensity in the recordings a and b occurred when the two mirrors were in exactly the same plane.

It should be noted that the coherence property of light defines the extent of a field's influence at a distance and not a quanta's size. The coherence of visible light can extend over a volume 10^{10} times larger than the estimated size of a quanta. The graphs in Figure 25 show axial coherence. Lateral coherence is a characteristic of light not normally considered. For example the lateral coherence of sunlight according to equation (17) is a mere 0.03 mm. This makes it very difficult to generate interference fringes with any form of lateral-sampling interferometer. Young, for example, in his double slit experiment was forced to use a pin-hole filter in order to obtain visible fringes.

Light from distance stars, however, has large lateral coherence when it reaches earth. Robert Hanbury Brown used a stellar interferometer in Narrabi, Australia, to measure the lateral intensity coherence of light from the star Betelgeuse, which he found to be approximately 10 feet. He then used the equivalent of equation (17) to calculate the apparent size of Betelgeuse. (ref 20)

A modified version of the Koesters prism (ref 21) was used as a starlight line-of-sight sensor in the Hubble Telescope Fine Guidance Sensor. The Koesters prism is a lateral-folding interferometer responding only to unsymmetrical wavefront aberrations such as tilt, due to a star being slightly off the telescope optical axis. It does not respond to "even" aberrations such as the notorious spherical error in the

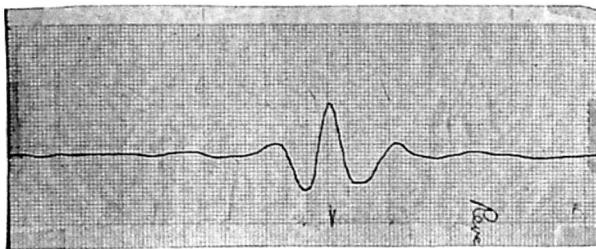

Figure 25a

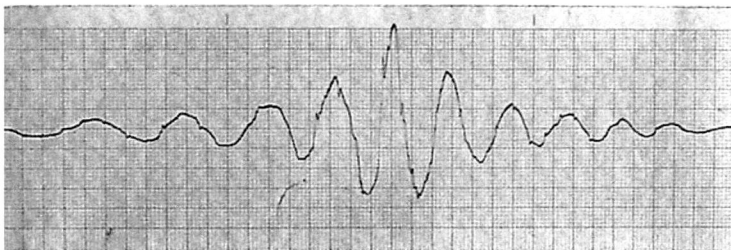

Figure 25b

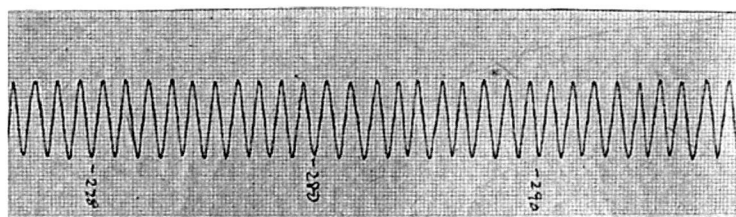

Figure 25c

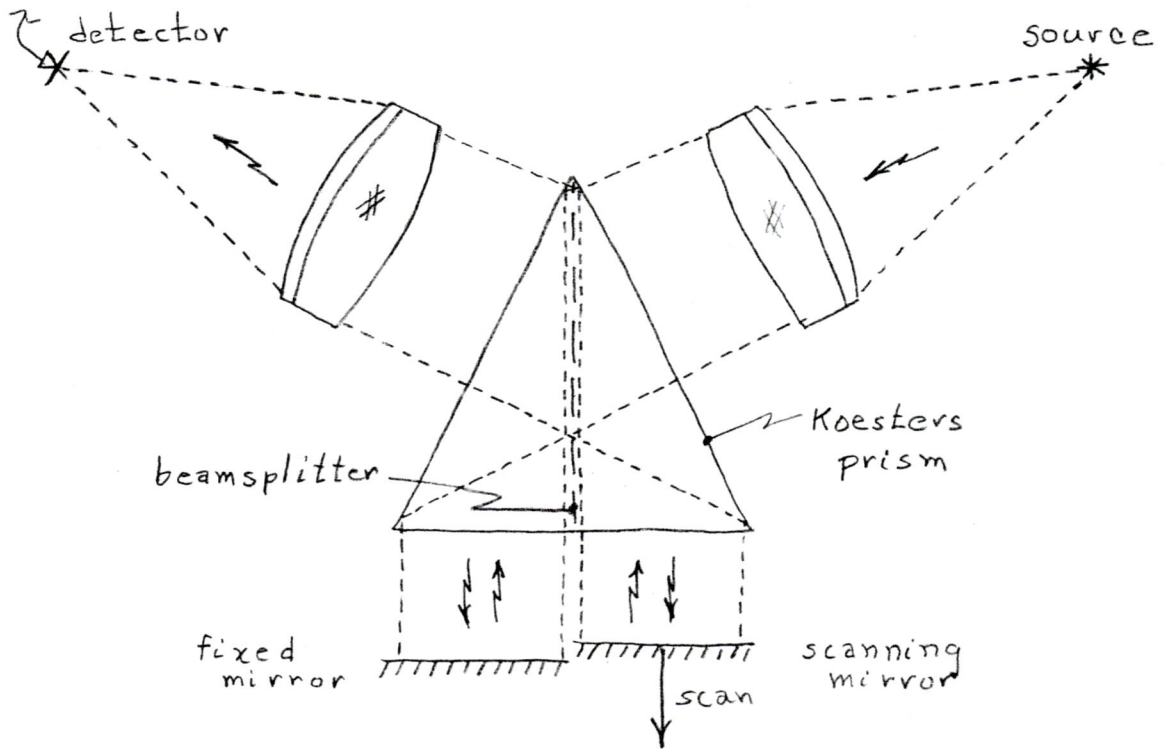

Figure 26

Figure 27

Hubble primary mirror. Thus the pointing capability of the Fine Guidance System was not disabled by the primary mirror error, as was reported in the media.

With careful control of the limiting aperture, a wavefront-folding interferometer can also be immune to Fraunhoffer diffraction! See Section 3.2.2.

This glorious picture, Figure 27, was assembled from Hubble data. (ref 22)

Experiment #5: Absorption Coherence

The term $\Delta\lambda$ in equation (16) is commonly understood to mean a narrow band of emitted light, i.e., a narrow wavelength band within which light exists. Experiment 5* demonstrated that, within a spectral continuum of light, a narrow band from which light has been removed will also generate interference fringes in an interferometer. Thus an absorption band of light can be said to be "a coherent lack of radiation." (ref 23) Figures 28 and 29 show Experiment 5 hardware and a portion of data from the experiment which demonstrated this phenomena. In this recording a positive deflection indicates the presence of coherent radiation in the scene. A negative deflection indicates the presence of coherent lack of radiation in the scene. The interferometer used for this experiment had an optical path difference and balanced interferometer-detector arrangement such that it did not respond to incoherent light.

Figure 28

* Experiment 5 was conducted by David Donovan and the author while both were employed by the Perkin Elmer Co., Wilton, Conn.

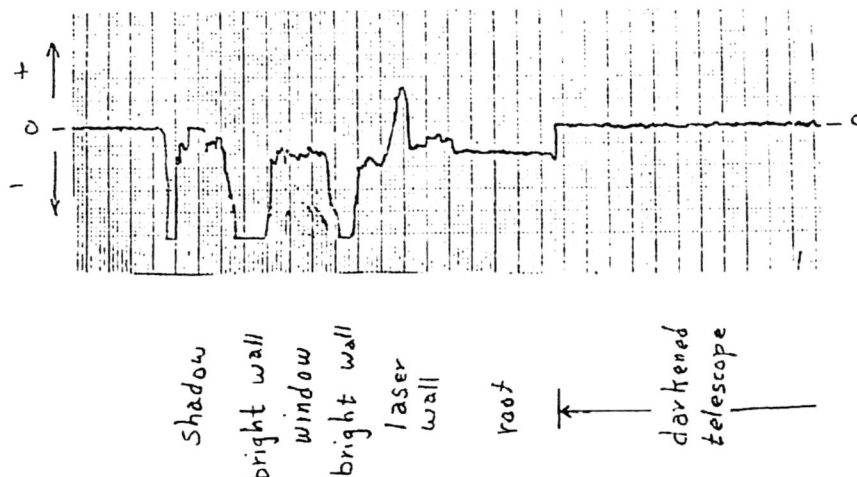

Figure 29

3. PROPAGATION EXPERIMENTS

3.1 Propagating Fields, Radio Wavelengths

The foregoing sections describe the field nature of light and how these fields are created. Experiments intended to synthesize individual fields identical to those of light were not successful, as reported in Section 2.2, therefore our quest for an understanding of light shifted to how light fields behave during propagation. Questions were raised such as "Does Maxwell's intuitive description of minute curling fields extend to propagating light ?", "Are the E and H fields of light permanently bound to each other?" and "What is the time-phase relationship between the two fields?" To answer these questions experimentally, it is necessary to probe propagating waves in the "far field." That is the subject of the following.

It was initially decided to investigate propagating fields at radio wavelength, where experimental techniques were available for the measurement of individual Et and Ht fields, subscript "t" indicating tangential fields. Far field radiation

from AM broadcast stations at distances between 10 and 65 miles were analyzed in Experiment 7 to determine the relative speed of propagation of the Ht and Et fields. Any difference in the relative speed would result in a difference in the relative time phase of the detected fields. Thus the phase response of individual components had to be calibrated.

3.1.1 Faraday's Law Of Induction

In Section 1.2 we discussed Faraday's early experiments involving the response of a coil of wire to a magnetic field. His Law of Induction, equation (3), describes this effect. As written, this equation says that the output will lag the input by 90 degrees. Experiment 6 was performed to demonstrate this fact.

Experiment #6: Faraday's Law

To obtain a pure magnetic field for Experiment 6, an adaptation of the rotating magnet assembly

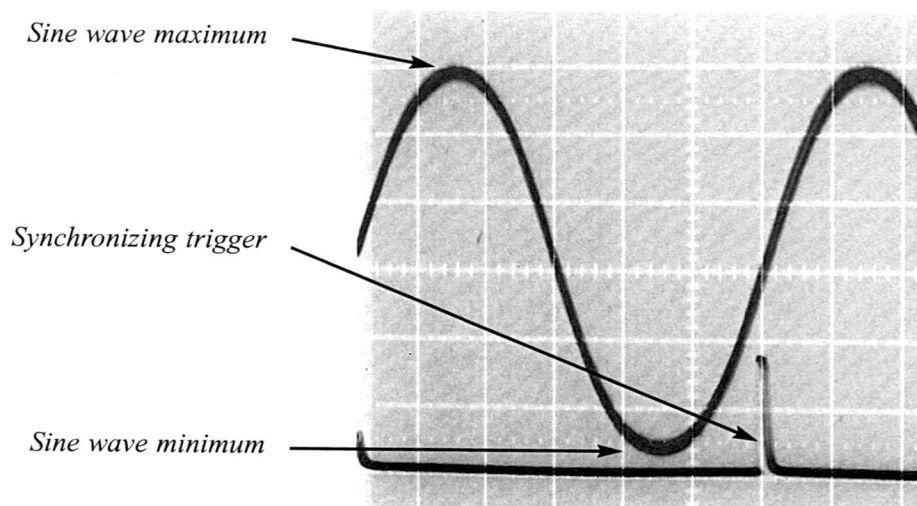

Sine wave maximum

Synchronizing trigger

Sine wave minimum

Figure 30

used in Experiment 1 was used. The lead weight drive was replaced by a sewing-machine motor and a slotted disk was added to generate a sync trigger for the oscilloscope display. This trigger occurred at the moment when the horizontal field flux was at it's maximum.

A photograph of the oscilloscope display is at the bottom of the previous page, Figure 30. Display of images such as this one have had their color edited to make a more printable picture. Note that in this figure the maximum positive value of the output occurs 1/4 cycle after the sync pulse. This corresponds to 90 degrees of time phase after the maximum of the horizontal field flux. Thus all measurements of E field phase delay, $\Delta\theta_a$ relative to the H field must be corrected by 90 degrees.

$$\Delta\theta_a = \Delta\theta_m - 90 \text{ degrees}$$

Where:

$\Delta\theta_a$ = *actual* time phase of E field relative to the H field

$\Delta\theta_m$ = *measured* phase of E field relative to the H field

3.1.2 Experimental Hardware and Test Procedures

Experiment #7: Field Phase Measurement

Two receivers were constructed, one responding only to the E field components of radio waves, and one responding only to the H field components of radio

waves. These receivers did not "detect" the energy. Rather they merely amplified the signal voltages intercepted by their respective antenna's and fed them to an oscilloscope for display. Each receiver consisted of an antenna plus a radio frequency amplifier. The old name for this type of receiver was TRF, or tuned radio frequency. Figure 31 contains a photograph of a TRF radio receiver circa 1920's which used vacuum-tube amplifiers. Operation of a TRF radio required tuning three narrow-band circuits to the transmitted radio frequency, sometimes a laborious procedure. In addition to being tuned to the same radio frequency, the two receivers must have identical phase response. A special phase calibration was incorporated into the experiment procedure to assure this response.

Figure 32 on the following page is a photograph of the complete Experiment 7 equipment assembly, located in the author's back yard, Southbury, CT. Figure 33 is a schematic

Figure 31

E field antenna

Oscilloscope photography hood

H field antenna

H field receiver

Figure 32

of the H field receiver, in which the antenna consisted of a 14" diameter coil, or loop, covered with grounded aluminum foil to eliminate response to E fields. Two E field antennas were used. The time-phase between the E and H fields was measured using a 7' vertical pole and a 15" diameter shorted loop. Curl experiments were made with the loop. Both the pole and loop received signals over a horizontal field of view of 360 degrees. This coverage was expected for the pole, but was a surprise for the flat loop. All amplifier parts were enclosed in sheet-iron boxes to eliminate spurious responses to E or H fields. Notice the oscilloscope in this picture has a cardboard hood for photographing the displayed E field and H field signals.

Experiment 7 tests were performed at variety of sights in the hills surrounding Southbury, CT. However best results were found right in the author's own back yard. It may be noted in various photographs that a majority of the experiments described in this book were performed either in the author's kitchen, living room or back yard, all courtesy of the author's wife.

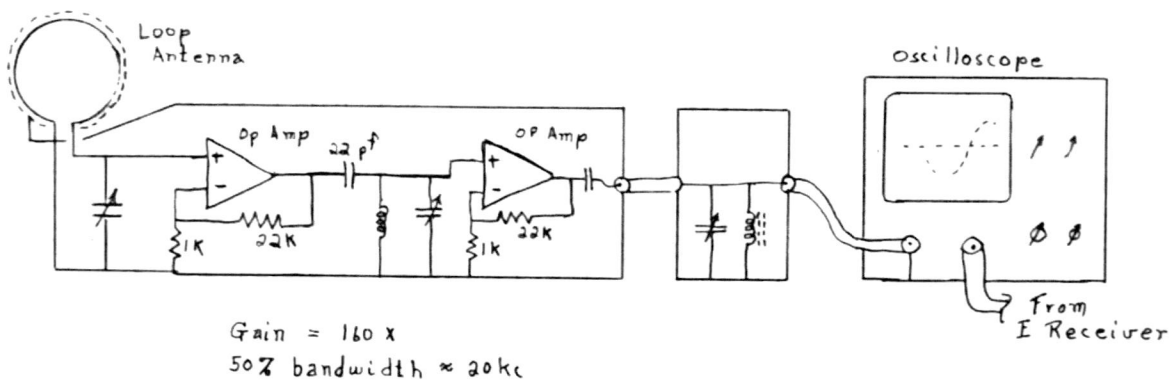

Figure 33

Receiver tuning used a Wave Generator which was calibrated by a crystal-controlled digital communication receiver, ICOM IC-R75, better known as a "ham radio." Fine tuning to produce identical phase, θ, response in the two receivers was achieved by feeding the same test signal through a high impedance to each antenna. Figure 34 is an example of the oscilloscope display from this procedure. This waveform is actually an overlay of two receiver outputs, showing a minimum phase difference between their responses. Accuracy of the time-phase calibration, $\in$, was found to be:

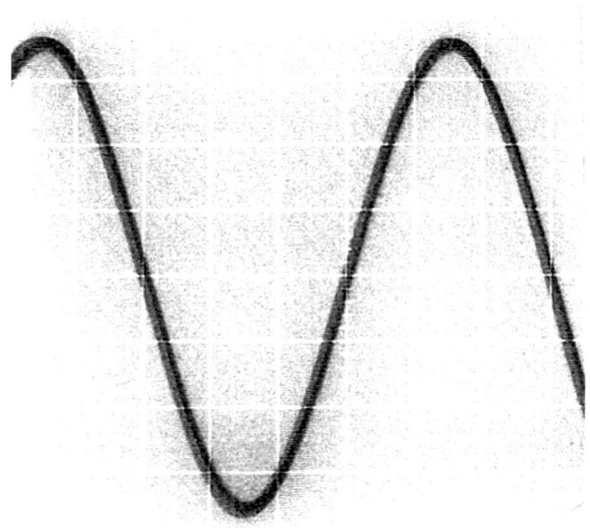

Figure 34

$$\in \ = 3 \text{ degrees peak}$$

The principle objective of Experiment 7 was to measure the relative time-phase, $\Delta\theta_a$ between the H and E fields. Figure 35 shows the method used to calculate $\Delta\theta_m$ from the display of signals from the two receivers. Both traces in this display are synchronized by the H field signal. Phase of the H field signal was identified by horizontal-axis crossings a, b and c. The positions of these crossings, as indicated by the "scale," were inserted into equation (18) on the following page. Traces such as this

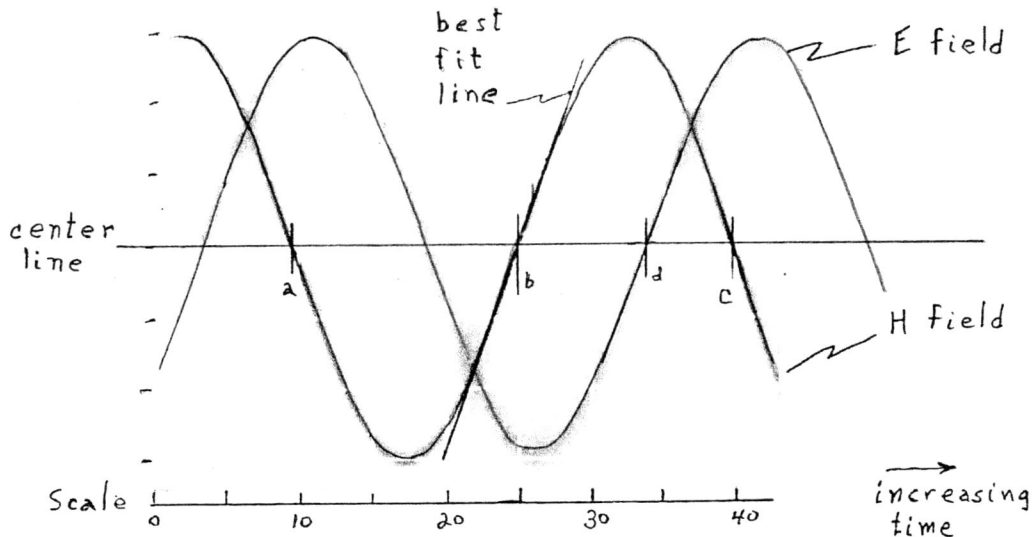

Figure 35

were recorded at moments when audio modulation was minimum.

Notice that in Figure 35 the display of "time" increases to the right. This results in point c occurring at a more recent time than point a. Also the positive-going crossing of the E field, d, occurs after the positive-going crossing of the H field. Or the E field is delayed relative to the H field. The time phase difference between the two wave forms, $\Delta\theta_m$, was calculated using the following:

$$\Delta\theta_m = ((b - d - \in) / (c - a)) \times 360$$
$$\text{(degrees) (18)}$$

Where:

b - d = time between E and H positive-going axis crossings

c - a = time between H waveform successive negative-going axis crossings, equal to one wavelength

In this figure the time-phase delay of the E field relative to the H field is:

$$\Delta\theta_m = -104.2 \text{ degrees}$$

A set of measurements made during a period of calm atmosphere gave an uncertainty of measuring time-phase differences of:

$$\sigma(\Delta\theta_m) = 0.2 \text{ degrees rms}$$

3.1.3 Field Phase Measurements

The objective of Experiment 7 was to measure the time-phase difference, $\Delta\theta_a$, between the E and H fields as a function of the time of day, weather and distance from the source. These were all far-field measurements. Radiation from a number of AM broadcast stations was used. Distances between the receivers and the broadcast antennas varied between 10 and 64 miles.

A typical set of measurements was made November 2004 at mid-day, when the atmosphere was relatively stable. Propagation measurements made at mid-day involve ground-wave propagation only. This eliminated multiple reflections from the ionosphere which would have thoroughly confused phase measurements. Figure 36 contains photographs of typical waveform displays for radio stations located at 10, 23, 41 and 62 miles away. Values of $\Delta\theta_m$ and of $\Delta\theta_a$ calculated using these wave forms and equation (18) are listed in Table 1 on the following page.

A first look at the values of $\Delta\theta_a$ in Table 1 do not tell us much. However, if the possibility of an additional 360 degrees or 720 degrees is considered for the longer distances then the numbers make sense. This is called wrap around, which can not be recognized in the $\Delta\theta_a$ numbers. Wrap around becomes obvious when the value of $\Delta\theta_a$ versus distance is plotted. Then the graph becomes a pretty picture, a picture as pure as a Puccini aria. See Figure 37 on pg. 38 for a graph of $\Delta\theta_a$ with this wrap around included.

E lags H

H

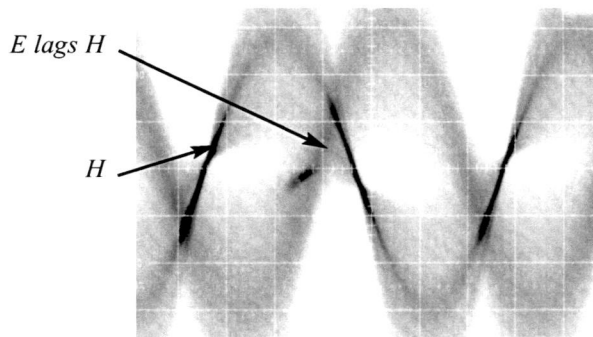

Fig. 36a.

E leads H

H

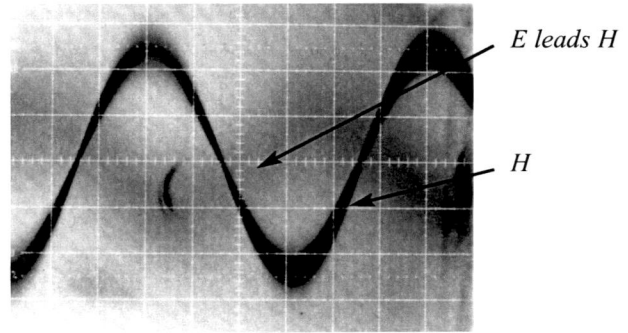

Fig. 36b.

E lags H

H

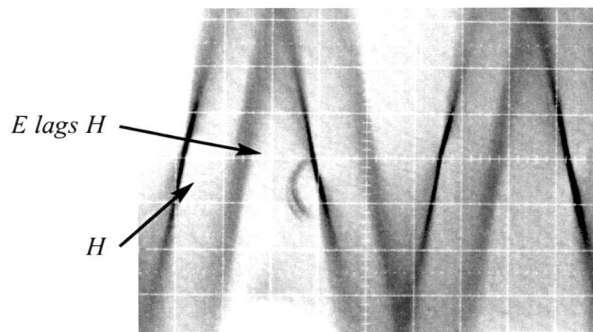

Fig. 36c.

E leads H

H

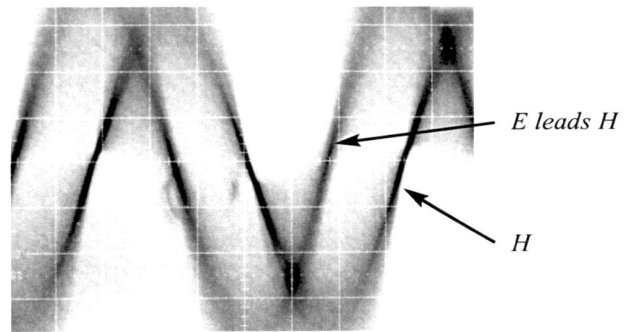

Fig. 36d.

Table 1
Values of E field phase relative to H field phase
Measured November 2, 2005
Southbury, Connecticut

AM Station	Distance (r)	$\Delta\theta_m$	$\Delta\theta_a$	Wrap Around	$\Delta\theta_a$ plus Wrap Around	Figure
WINE	12	-131	-221	0	-221	Fig 36a
WELI	23	+131	+41	-360	-319	Fig 36b
WTIC	41	-78	-168	-360	-528	Fig 36c
WINS	79	+90	0	-720	-720	not shown
WCBS	69	+86	-4	-720	-724	Fig 36d

Figure 37 graphs the phase delay between E and H fields versus the distance to the radio transmitter. Linear regression analysis of the first three data points in this graph showed:

Slope = 10 3/4 degrees/mile

Statistical uncertainty, γ = 0.97 degrees rms

When r = 0, then $\Delta\theta_a$ ~ -90 degrees

This last number compares favorably with the predictions contained in Figure 7 in Section 2.1.1, i.e. that a pole antenna emission starts with the curl E field lagging behind the curl H field by 90 degrees.

It can be concluded from Figure 37 that during ground wave propagation, the E field travels through a normal atmosphere at a velocity of 11 degrees of time phase per mile slower than the H field.

Evidence from a scattering of measurements in different seasons indicates that variations in the slope of the graphs of phase delay are probably due to ground moisture variations. Ground

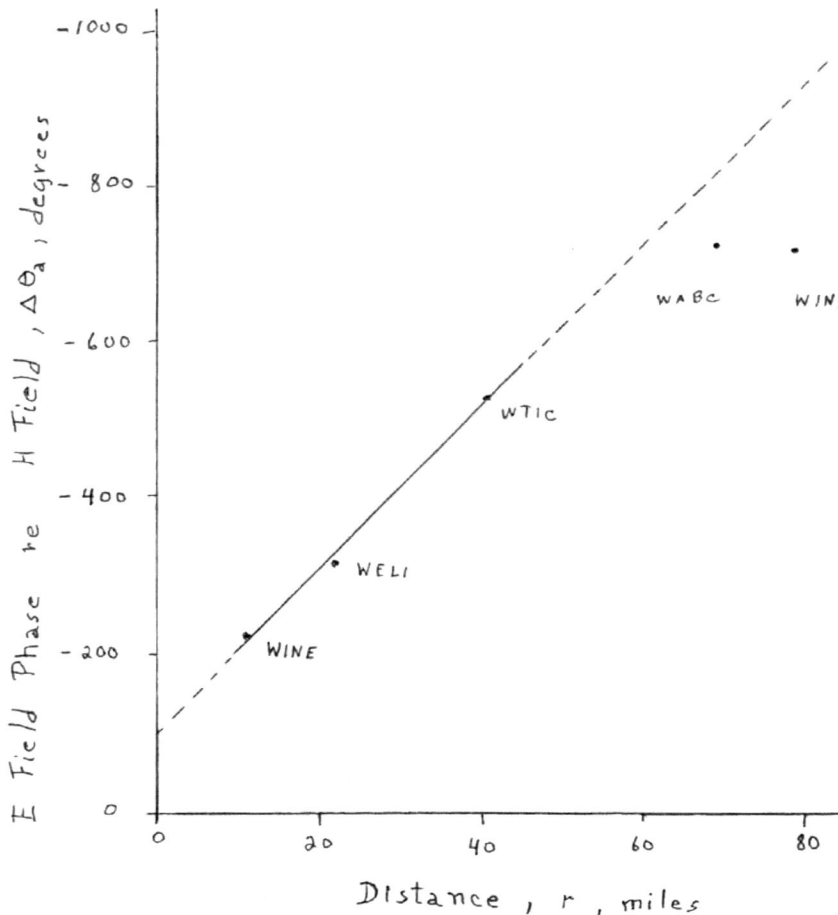

Figure 37

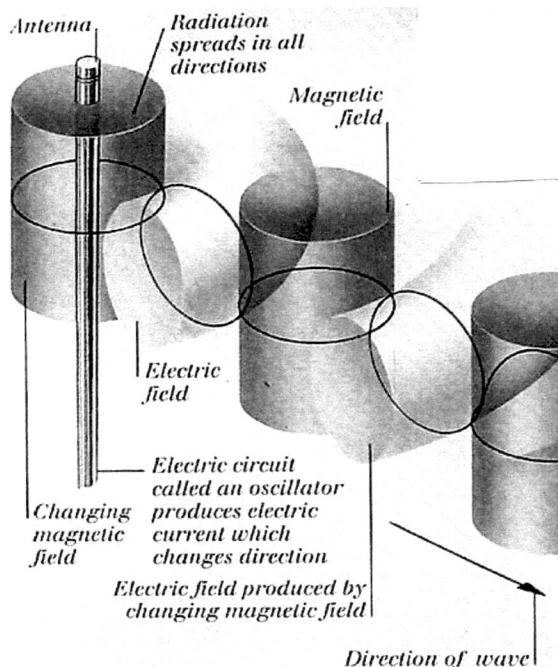

Antenna
Radiation spreads in all directions
Magnetic field
Electric field
Electric circuit called an oscillator produces electric current which changes direction
Changing magnetic field
Electric field produced by changing magnetic field
Direction of wave

Figure 38

moisture introduces a small increase in $\in$, which reduces the ground wave velocity per equation 6e.

It was also learned that when propagation passed above high density urban areas the phase difference between E and H beams was considerable reduced. This is demonstrated by the data from WABC and WINS in Figure 37. WABC transmitter was located in the Bronx bureau of New York City and WINS transmitter was located in New Jersey, across the Hudson River from Manhattan, NY.

Since the E field travels at a different velocity than the H field, it can also be concluded that radio wavelength E and H fields propagate without mutual "curl." This implies that Maxwell's equations describe the origin of electromagnetic radio waves but <u>not</u> necessarily their propagation.

Figure 38, from a popular physics picture book (ref 24) shows that the initial phase of the two curling fields agree with the predictions made in Figures 6 and 7.

> *A physicist should investigate, challenge and unceasingly seek the joy of uncovering Nature's Laws.*

3.1.4 "Curl" Experiment

Experiment #8: "Curl" Experiment

Experiment 7 showed that propagating radio waves consist of cyclic E and H fields. It also showed that the two fields propagate at slightly different speeds through an atmosphere. This implies that there is no interaction between the two fields in far-field propagation. By inter-action we mean curl H, curl E, curl H, etc, as described by Maxwell's equations (6c) and (6d). With timidity, one could inquire if Maxwell's "curl" ever exists. The purpose of Experiment 8 was to demonstrate that they do exist in the near field, i.e. short distances compared to the emission wavelength.

The first step of Experiment 8 was a calculation of the theoretical time phase between E and H fields in a propagating wave. Terms E and H refer to the tangential fields of propagating electromagnet waves. The pertinent equations used in the calculation have been re-arranged into a cause and effect form in Table 2. Figure 39 shows the time phase relationship

Table 2

Equation			Based on	Page	Figure
$+\dfrac{\partial E}{\partial t}$	$\Rightarrow$	$+\ \text{curl}\ H$	*(6d)*	*6*	*Maxwell*
$-\dfrac{\partial H}{\partial t}$	$\Rightarrow$	$+\ \text{curl}\ E$	*(6c)*	*6*	*Maxwell*
q	$\Rightarrow$	$E\ \text{field}$	*(4)*	*5*	*Coulumb*

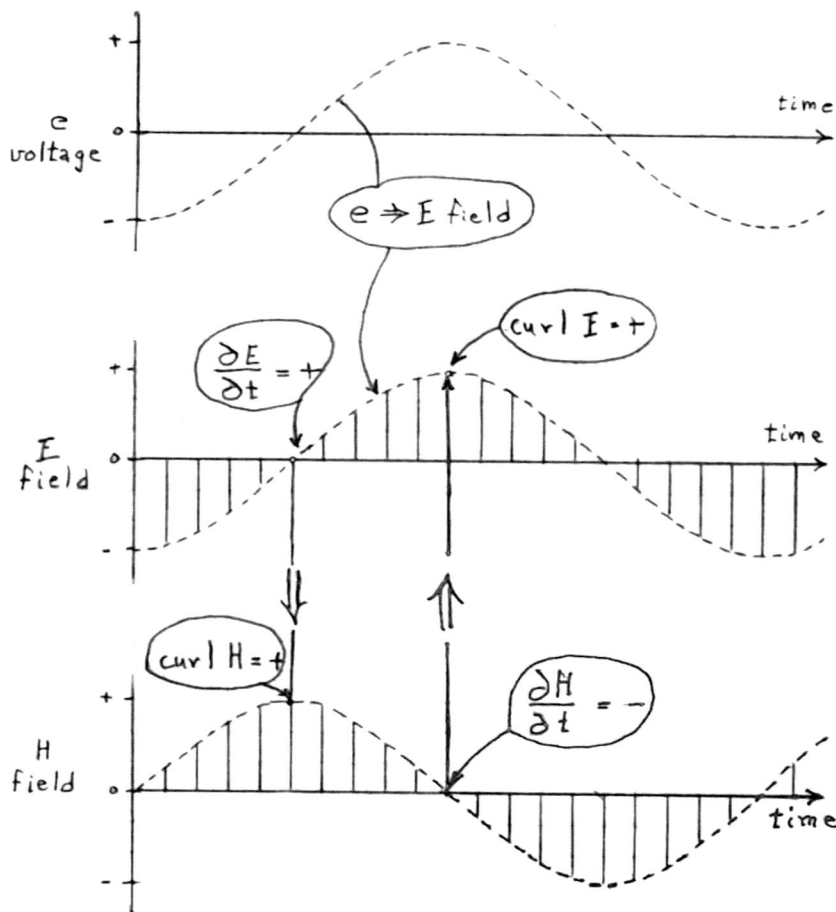

Figure 39

between E and H fields in accordance with Table 2. Note that the curl E field lags the H field and that the curl H field leads the E field. The reader may note a similarity between Figure 7 and Figure 39. In the latter figure the sequence starts with applied voltage to agree with the experimental arrangement.

Figures 40 and 41, on the following page, picture the instrument arrangement for both parts of the experiment in which a radiated field created a companion field via "curl"

action. E and H receivers were the same ones used in Experiment 7. The signal generator was set at 1,000 kc, or 1.0 mc.

The following is the calculated phase difference for Figure 40. The "curl" H field led the E field by + 90 degrees, per Figure 39. The H receiver loop antenna had a phase delay of -90 degrees, per Faraday's Law. The calculated E signal phase, relative to the H signal = 0 degrees. Therefore, the E and H fields have zero degree phase difference.

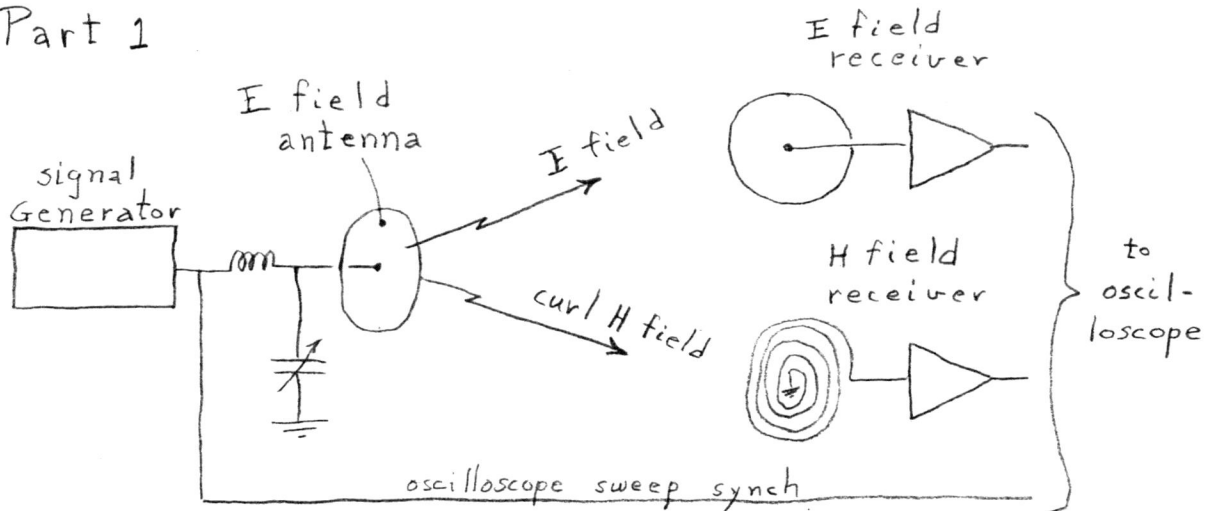

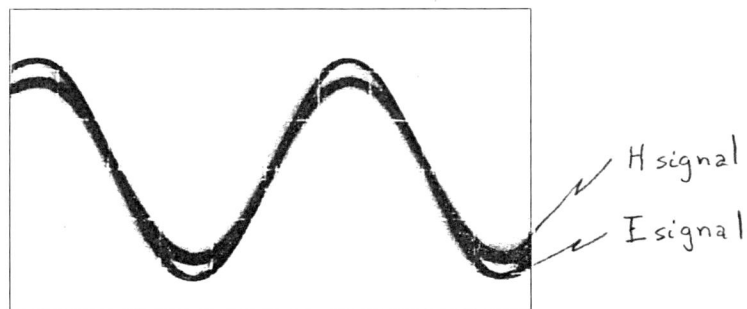

Figure 40

The following is the calculated phase difference for Figure 41. The "curl" E field lagged the H field by - 90 degrees, per Figure 39. The H had a receiver loop antenna phase delay of - 90 degrees, per Faraday's Law. The calculated E signal relative to the H signal phase = - 180 degrees. Therefore, the E and H fields had 180 degrees phase difference.

3.1.5 Observation

Thus it can be said that in the radio portion of the electromagnetic spectrum the "curl" action implied by Maxwell's equations is real in the near field. It can also be observed that at the emission of radio-wavelength energy the E and the H fields are separated in time phase by 90 degrees.

The absence of "curl" in the far field, Experiment 7, and the presence of "curl" in the near field, Experiment 8, could be accounted for if one assumes that "curl" action occurs only in the presence of electrons.

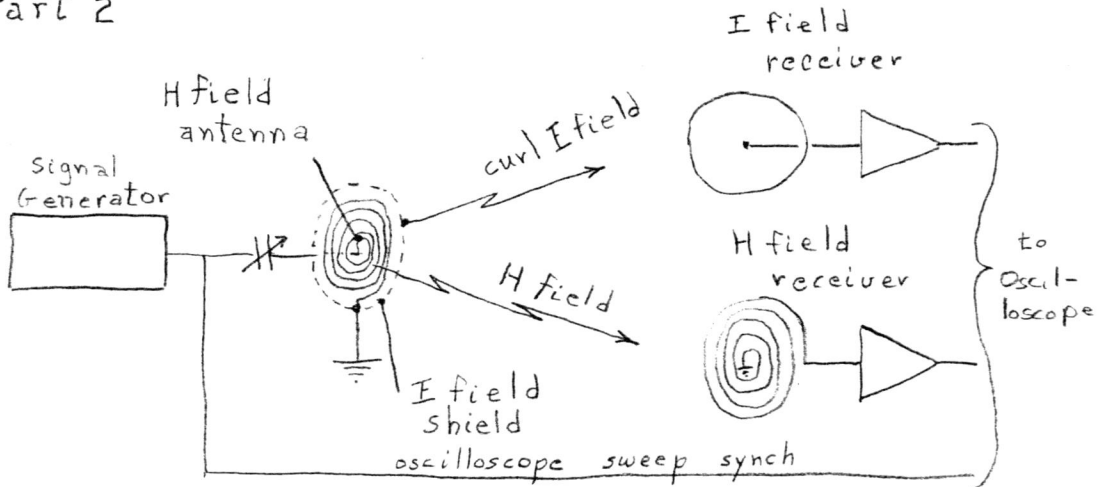

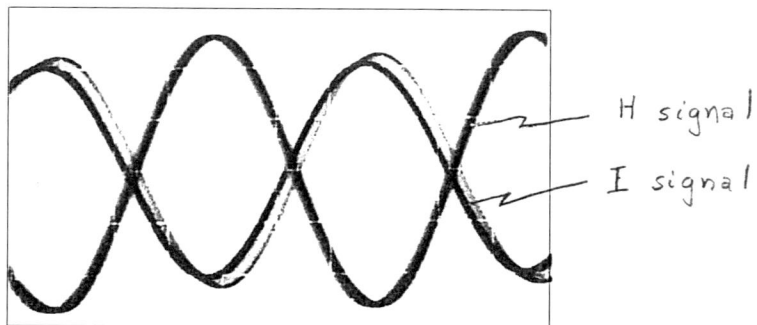

Figure 41

3.2 Propagating Fields, Visible Band Wavelengths

3.2.1 Two -Field Viewpoint

Bohr showed that an oscillating electric charge, i.e. an electron, can generate electromagnetic energy, see Section 2.1.2.2. Planck theorized that the emission of electromagnetic energy is not a continuous stream of energy, but rather it is constrained to occur only in many discrete elements of energy, see Section 2.1.2.1. Then Einstein demonstrated that the detection of electromagnetic energy also occurs only in many discrete elements of energy, called "quanta," see Section 4.3. This might lead one to believe that the propagation of electromagnetic energy between emission and detection also occurs only in discrete elements of energy, called "photons." <u>Not</u> necessarily so.

Consider what is propagating. Experiments such as described in Section 3.1.3 have demonstrated that propagating electromagnetic energy consists of propagating independent E and H fields. But what are E and H fields? Over 140 years ago Faraday called them "induction." Today we still don't know what they are, but we do know what they do and we call them "influence," caused by an electron at a distance. Think a moment about a propagating electric or magnetic influence. It has no gravity and there is nothing bounding it's size, large or small, not even Planck's constraint, which applies only at the origin and detection of light.

We have concluded that: "Propagating electromagnetic energy, radio or visible

wavelengths, consists of independent and unconstrained electric and magnetic fields." When considered from the "two-field viewpoint," a number of intriguing ideas come to mind. First, the single-photon version of Young's double-slit experiment need no longer be a mystery, Section 3.2.3. Second, there is a real possibility of eliminating scintillation of ground based starlight observations, Section 4.6. Third, what would an independent magnetic field camera reveal about our world? The list goes on, but we have time for only a few experiments. It is up to younger physicists to adopt this viewpoint and open new doors to Nature's Laws.

3.2.2 Diffraction Experiments

Experiment #9a: Diffraction Experiments

What might be considered a classic Fizzeau diffraction experiment in Figure 42 on the following page is from the laboratory session of a physics class the author attended during his first year of college, 1940. This is a recording on film of a diffraction pattern formed by light which had passed through a very narrow vertical slit. If light were to have passed through the slit and arrive at the film plane undisturbed it would have made an exposed pattern with infinitely sharp edges. But as can be seen, this is not the case. There is a pattern of exposure level at the two edges with intensity varying in a sin wave manner perpendicular to the edges. Diffraction is a lossless event. The only lossless effect known to generate a sin wave pattern is interference. Light can

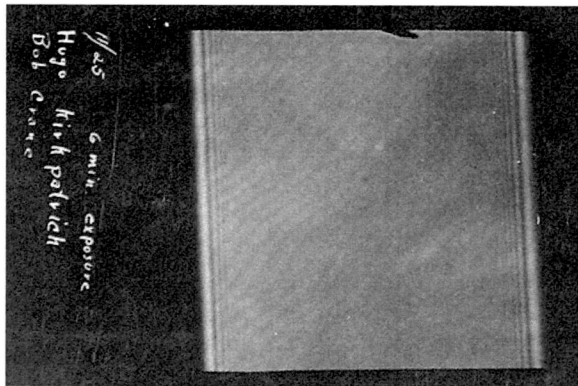

Figure 42

not interfere with itself. Therefore, there must have been at least two wave fronts impinging on the film, one from straight through transmission and one emitted from each of the two slit edges.

It is well known that this diffraction pattern can be calculated mathematically using

Huygen's theory. The pattern can also be computed using Cornu's spiral. (ref 25) Both true, but this should not distract from the fact that the sin wave pattern of constructive and destructive intensity requires the presence of two wave forms. This means that the opportunity for scientific adventure lies not in pattern analysis but rather in the question, "Why are there two wave fronts?" or "What secret of Nature generates the second wavefront?."

Experiment #9b: Diffraction Experiments

What has been called Fizzeau diffraction occurs at a slit edge, or a knife edge. The second diffraction experiment examined Fizzeau diffraction with a small coherent beam, < 1 mm in diameter. This allowed a major part of the diffracted beam to be seen

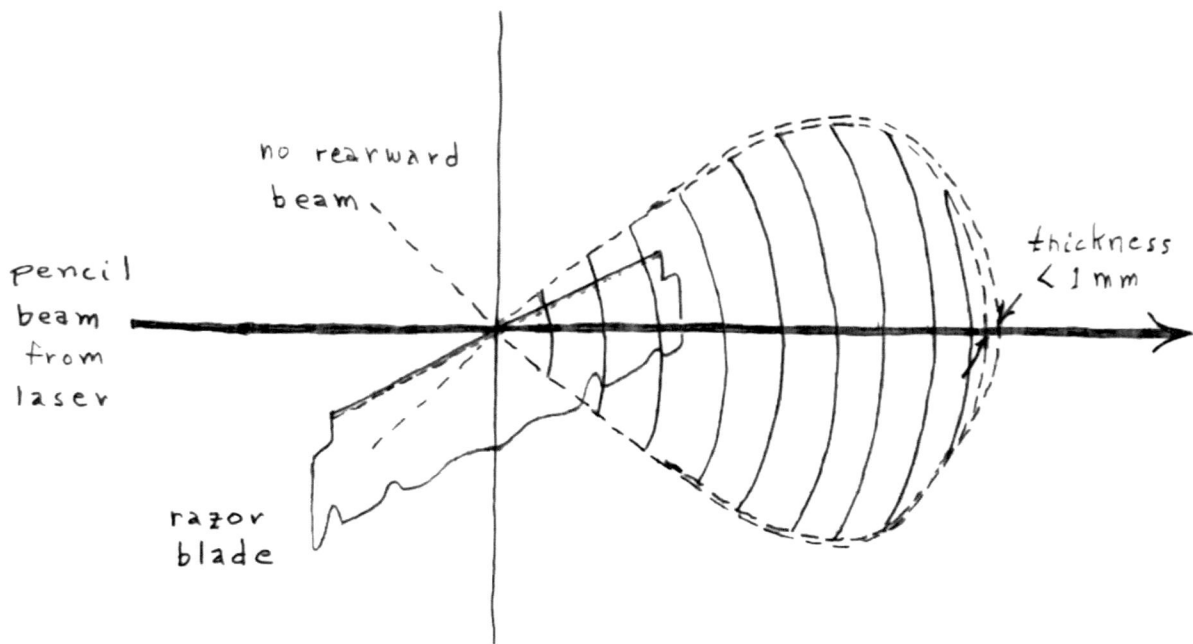

Figure 43

without interference. A sketch of this "second beam," as we will call it, in Figure 43 shows that it is a thin, fan-shaped beam, which is parallel to the straight through beam, perpendicular to the knife edge, and expands on both sides of the straight through beam. The second beam has a parabolic wavefront as it expands and is coherent with the straight through beam.

Figure 43 shows that the second beam appears only in the forward direction, there is no backward beam. However, there are backward reflections, both specular and diffuse.

Figure 44 is a photograph of a collimated laser beam after it had passed by a razor blade. In this case the film was located 1 3/4 inch from the razor blade, with no optical element between the two. Analysis of this fringe pattern showed that the second beam wavefront has a parabolic shape, and not the anticipated circular shape. Note the limited extent of these fringes.

The second beam has been found to originate from both conducting iron and non-conducting glass. This suggests that it is emitted by a valence electron, rather than from a free conducting electron. Intensity of the second beam is also a function of the conductivity of aperture edge material. An experimental comparison was made of two second beams, one from a stainless steel edge and one from a rubber edge which had been saturated with absorbing ink. The steel edge produced a very large second beam compared to that from the rubber.

Figure 44

It may be concluded that the second beam is generated by forward scatter or a form of "stimulated emission" from an electron. Conditions for the stimulated emission are not present in this experiment. However our search for an answer will continue.

Experiment #9c: Diffraction Experiments

Classic physics text books imply that in a diffraction-limited optical system, diffraction at a limiting aperture causes the image to be smeared. When fringes due to a diffracted beam are examined close to the aperture, as in Figure 44, they are called Fizzeau fringes. When they are examined at a distance from the aperture they are called Fraunhoffer fringes. In both cases the fringes are accounted for by mathematical derivations.

We have suggested in a previous paragraph that fringes close to an aperture are due to a "second beam" which is generated at the aperture edge. As a collimated beam propagates away from a limiting aperture, fringes will appear along the edge of the beam. Instead of Fraunhoffer fringes, we will henceforth describe these fringes as due to a "boundary beam." The remainder of this section will describe measured characteristics of this boundary beam.

Figure 45 is a sketch of the geometry for the second and boundary beams. Both beams generate recognizable fringes by interference with the straight through beam, and both beams have parabolic wave fronts. Thus the boundary beam is most likely an extension of the second beam.

Most interestingly, it appears that the boundary beam grows in size as it propagates away from the aperture. Figure 46 on the following page contains a series of photographs of

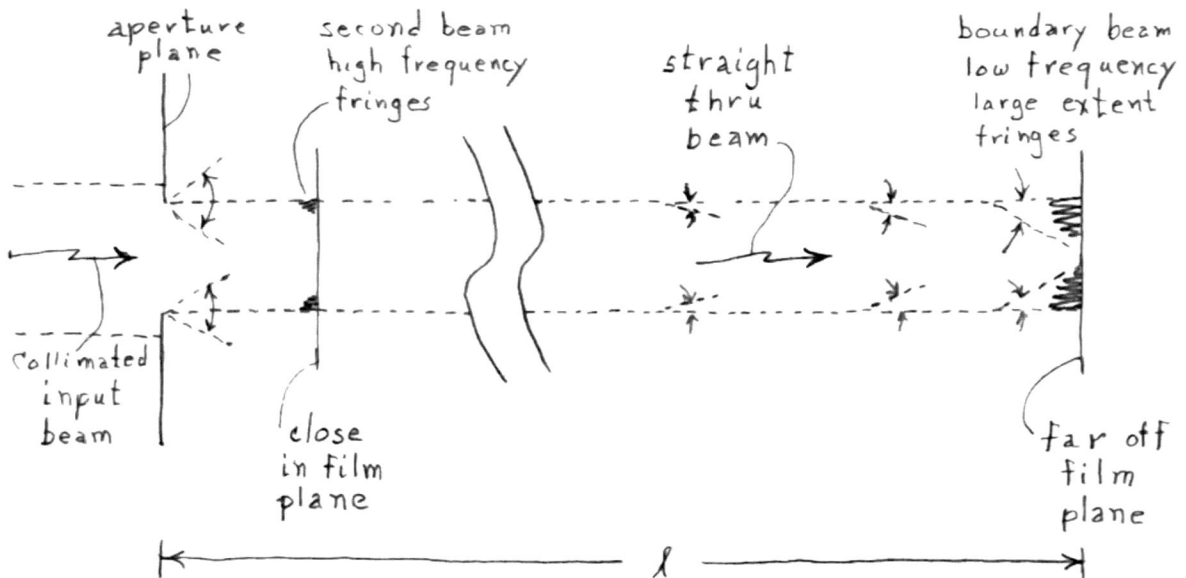

Figure 45

a 1 inch diameter coherent beam as a function of the distance, *l*, from a limiting aperture. Our interpretation of this growth says that there must be amplification of the boundary beam as a function of *l*. Once again the evidence points to the possibility of a form of stimulated emission.

The idea that stimulated emission could occur in free air is pretty far out, especially in view of the fact that only electrons emit light.

Figure 47 on the following page is a sketch of the apparatus being used to search for a way to bring this idea back to reality.

Stimulated emission was explored by Charles Townes in extensive spectroscopy experiments which led to his discovery of the concept for a maser, a microwave amplifier. This was followed by his invention of the concept for the laser, light amplification by stimulated emission of

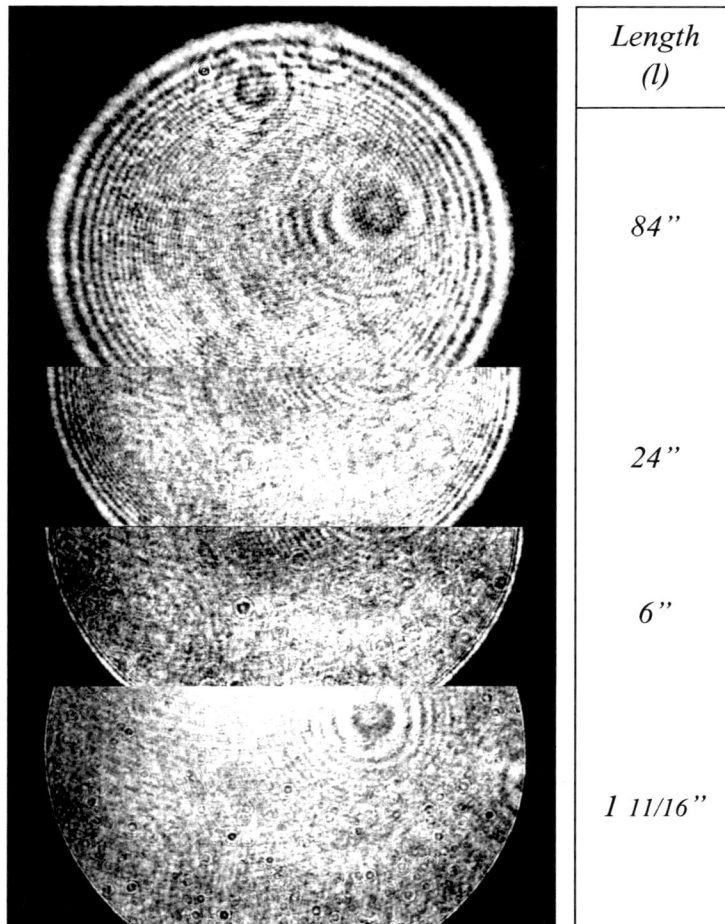

Length (l)
84"
24"
6"
1 11/16"

Figure 46

radiation. (ref 26) For stimulated emission to provide gain, there must be an "inversion" of the natural ratio of stable to excited valence electrons. Is it possible that the straight through beam is strong enough to create this inversion?

This is an "on-going" experiment, and should be continued with the objective of learning where the energy comes from to support fringe increase as *l* increases.

The importance of aperture diffraction becomes apparent when the collimated beam of Figure 47 is brought to focus by an imaging lens. The resulting image, called a "point spread function," is not a single dot, but rather it becomes a central disc surrounded by a set of rings. The disk and set of rings are due to the energy in the boundary beam. The rings increase as exposure increases. This is demonstrated by the series of exposures near the bottom of the front cover. Reduction or even elimination of limiting diffraction would indeed be fun!

There are two nifty equations relating the size of the point spread function to the geometry of a telescope.

$$\delta_{focal} = \left(\frac{f}{\#} \right) \times \lambda \qquad (19)$$

$$\delta_{space} = \frac{\lambda}{D} \ (radian) \qquad (20)$$

Where:

D	=	aperture diameter (same)
f l	=	focal length (dimensions)
f/#	=	D / f l
γ	=	size of point-spread function
λ	=	wavelength

The term nifty refers to the fact that these equations are not rigorously exact, but are simple and easier to use where exact numbers are not required.

> *"The way to progress is never swift or easy."*
>
> *- Marie Curie*

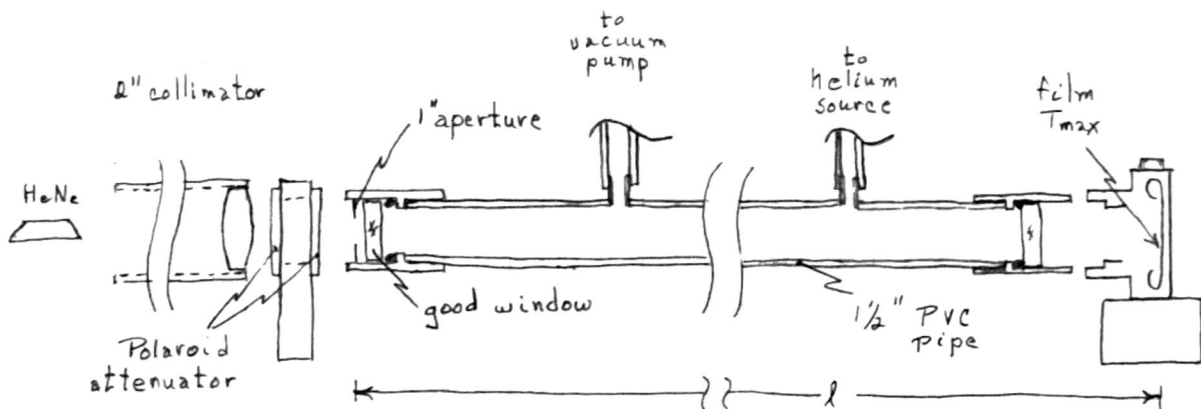

Figure 47

3.2.3 Young's Two-Slit Interferometer

About 200 years ago Thomas Young's two-slit experiment made history because it demonstrated the idea of interference, which is based upon the wave theory of light. (ref 27) The two-slit interferometer made history when it was performed with very low input intensity. The intensity was reduced to the level that only one photon was detected at any one time. Yet the interference pattern formed by one photon at a time was identical to that obtained with higher light level input. (ref 28)

A quandary arose from the assumption by experimenters that light propagates as photons, with each containing one quanta of energy. We have shown that this is not true. Light propagates as E and H fields.

Fields can be added or divided with no limit on the energy they contain. Remember, Planck's constraint is a characteristic of electrons involved in the emission and detection of light, and not a characteristic of propagating light.

Thus "single photon" interference experiments very likely involved field division, resulting in two coherent fields, each containing 1/2 quanta of energy. If one field passed through each slit, an interference pattern would have been generated when they combined at the fringe plane. In addition the combined fields would contain a total energy at the fringe peaks of 1.0 quanta, sufficient to be detected by an electron and counted as one photon. Simple.

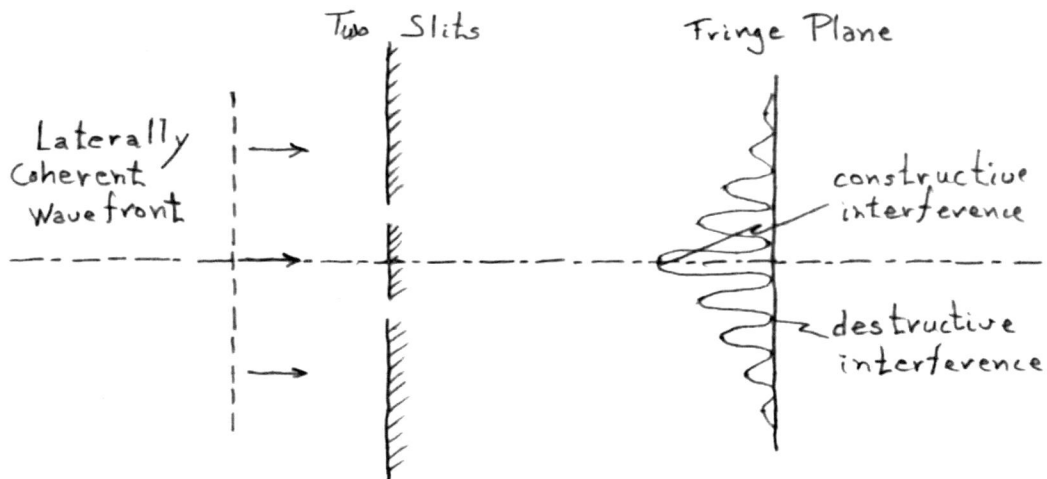

Figure 48

4. RECEPTION

4.1 Detection Observations

The following discussion of electromagnetic wave detection is divided into two parts, a radio frequency band and a light wavelength band. Figure 49 below shows these bands to be separated at 300 Ghz or 1000 micron. This is the highest frequency at which radio frequency receiver local oscillators can be manufactured.

In the radio frequency band, "detection" can be divided into five separate functions:

1. Field draw down, either E field or H field

2. Absorption of many E field quanta by many electrons, or absorption of H field energy by a loop or coil of wire

3. Detection to separate data from a radio frequency carrier

4. Frequency filtering to isolate data

5. Data processing ⇒ information

In the light wavelength band, "detection" can be divided into similar functions:

1. Field draw down, E field

2. Single quanta absorption draw down of E field energy by an electron

3. Electrons current (carrying data), chemical reactions, biological processes or heat

4. Frequency filtering for a variety of purposes

5. Data ⇒ information

Field draw down is important to both bands so it will be described in the following Section 4.2.

Single quanta absorption by an electron was discovered by Einstein while experimenting with Photo Emission (ref 28), see Section 4.3. In addition to Einstein's pioneering work, function 2 above is still a source of many questions. For example, "Why does an electron instantly absorb exactly one quanta, as limited by Planck's constraint?", see Section 4.4 or "Why

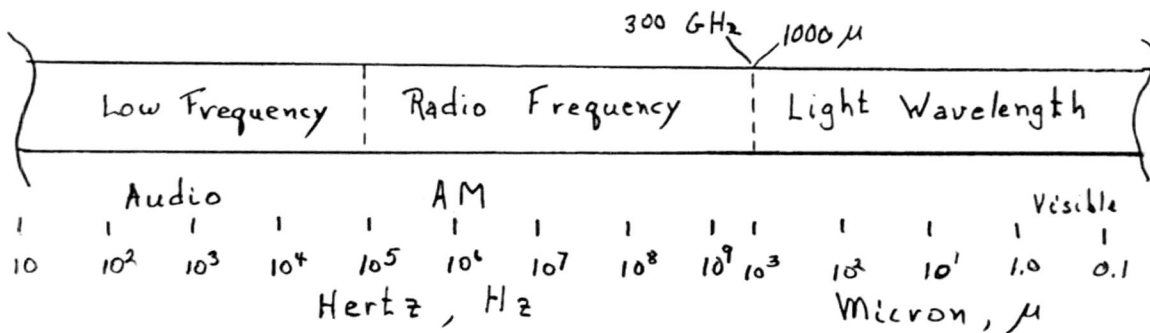

Figure 49

does an electron respond to only the E field half of light?"

To the first of the above questions all we have to offer is, "That's the way electrons are." However, the second question once again offers an opportunity for scientific adventure. We have inverted the question into a search for a detector which responds to the H field half of light, Section 4.5.2.

Can you imagine what our surroundings would look like to an H field detecting camera?

4.2 Field Draw Down

Experiment #10: E Field Draw Down

Numerous examples have been sited earlier in which a relatively small receiving antenna has extracted energy from an extended E or H field. Experiment 10 is a demonstration of E field "field draw down." A photograph of the experimental set up will be found on the following page. The emitter consisted of a signal generator feeding a single ball radiator. The receiver combined another single ball as an antenna, 18 mm diameter, feeding an integrated circuit op amp. The op amp output went directly to an oscilloscope. Note that the estimated time constant, τ, of the receiver input is:

$$\tau = 45 \; \mu\text{sec}$$

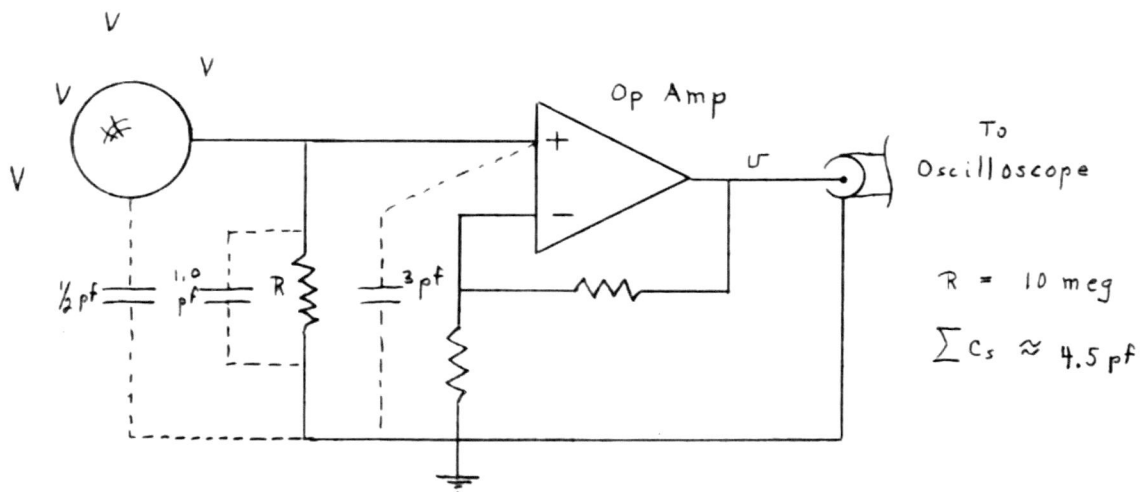

Figure 50

This experiment was performed with square-wave voltage, V, at 300 and 3200 cps and with distances, r, of 10 cm and 63 cm with the same results. There is ample evidence showing that the results are also valid up to microwave frequencies. It is the author's belief that they are equally valid at light frequencies.

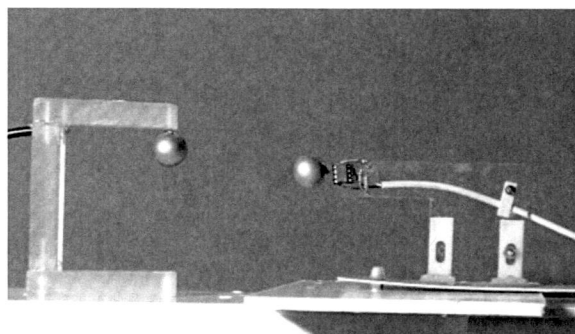

Figure 51

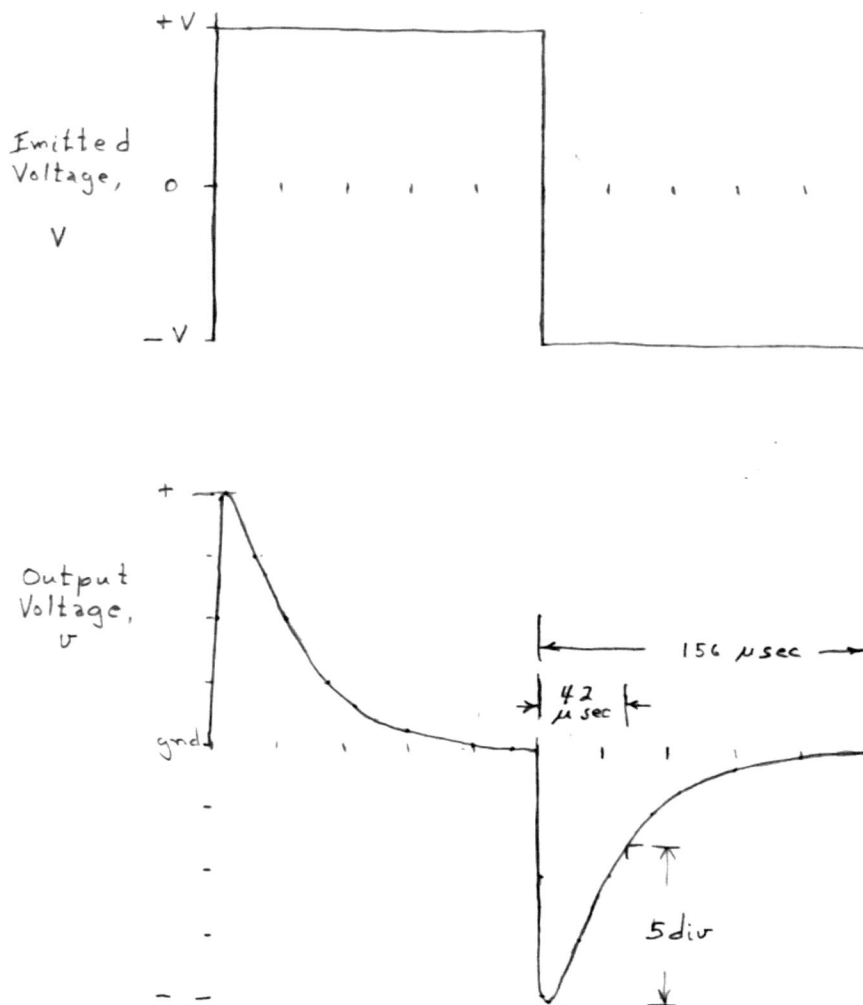

Emitted Voltage, V

Figure 52

The sequence of Experiment 10 started with a square-wave modulated field, V, at the emitter. This field then propagated to the receiver ball antenna. The ball and associated stray capacity instantly charge up to voltage v which was equal to the field voltage, V. Voltage v then discharged to ground through resistor R, bringing field voltage V with it. This caused a gradient in the field surrounding the ball. This is what I have labeled "field draw down."

Wave forms were drawn from the oscilloscope display of the emitted waveform, V, and the op amp output, v. Note that the decay of the measured voltage has a time constant of 42 µsec which compares favorably with the estimated 45 µsec for the input circuit. This implies that the field draw down was due to the receiver ball and its input circuit.

4.3 Einstein's Photo Emission

Einstein thought of light in terms of a stream of individual "light quants," each one of which carried energy equal to Plank's hv. He discovered and described the detection of these "quants" by an electron. His experiments involved photo emission, which is the escape from a conducting surface of a valence electron which had been excited by the absorption of one quanta of energy, defined by equation (11).

The key element of his experiment was the use of an adjustable dc voltage. The value of that voltage which just prevented a photoelectron's escape from the surface of the carrier

metal, allowed the calculation of the electrons kinetic energy, which in turn was equal to Planck's quanta. Einstein's photoemission equation may be written:

$$h v = (mv^2/_2) + \phi \text{ (joule)} \quad (21)$$

Where:
v = frequency of input light (1/s)
hv = Planck's estimate for the minimum energy of a light quantum (joule)
$mv^2/_2$ = kinetic energy carried by the photo electron emitted from an atom (joule)
ϕ = energy required to overcome the electron binding force at the metal surface (joule)

Thus Einstein's experiments were empirical confirmation of Planck's prediction of the energy carried by one "light quant," subsequently called one photon. This understanding of light detection, combined with Bohr's model of the atom for light generation by a bound electron, launched the era of quantum physics.

What we consider to be normal detection of light is a simple case of equation (21), in which hv is insufficient to overcome ϕ, in which case the freed valence electron remains within the metal to increase it's temperature or add to its conduction electrons.

There is a second subtlety hidden in the photo detection phenomena. This involves a question that is common to most situations where electromagnetic energy is detected. The question is: "How does a small detecting element, such as an electron, extract energy from an extended electromagnetic field?" As an example, in photo-detection this question is posed by two numbers:

1. Arguments presented in Section 2.1.2.3, Birth of a Photon, lead to an estimate for the extent of a quanta's E field to be 5×10^{-11} m

2. An electron has been estimated to have a size on the order of 5×10^{-15} m.

We believe that this discrepancy can be explained by a combination of coherence, described in Section 2.3.3, field draw down, Section 4.2, and Einstein's photo detection, Section 4.3.

> *"The nature of light is a subject of no material importance to the concerns of life or to the practice of the arts, but it is in many other respects extremely interesting."*
>
> *-Thomas Young (1773-1829)*

4.4 Electron Absorption

Experiment #11: Electron Absorption

The objective of Experiment 11 was to demonstrate the "all or none" behavior of an electron when absorbing light-wavelength quanta. The reason "why" this behavior occurs is known only to Alice, and she won't tell.

Figure 53 is a photo of the apparatus used for Experiment 11. This consisted of two white-light sources, both of which illuminated a single PIN diode detector. One source, referred to as the "test" source, was modulated by a

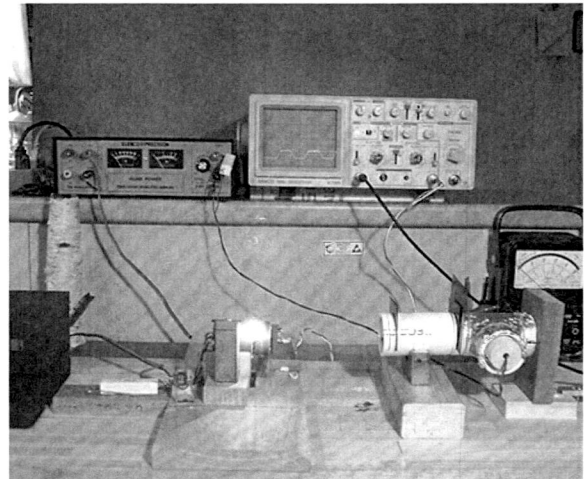

Figure 53

rotating disk with spaced fingers. This generated a square-wave modulation. The second source, referred to as a "bias" source, had adjustable intensity but was not modulated. Broad-band incoherent sources were used to avoid spurious interference effects.

The following is a layman's understanding of what happened inside the PIN when the two beams illuminated it. Quanta from the test source were absorbed by electrons, one quanta per electron. The author believes that these are valance electrons, i.e. those bound in a an atom's outer orbit. If a quanta is sufficiently energetic the electron escapes from the valance bond and becomes a free electron available for conduction. A small voltage swept the electrons out of the PIN. These test electrons passed through a "load" resister, generating a test voltage which was displayed as a square wave by an oscilloscope.

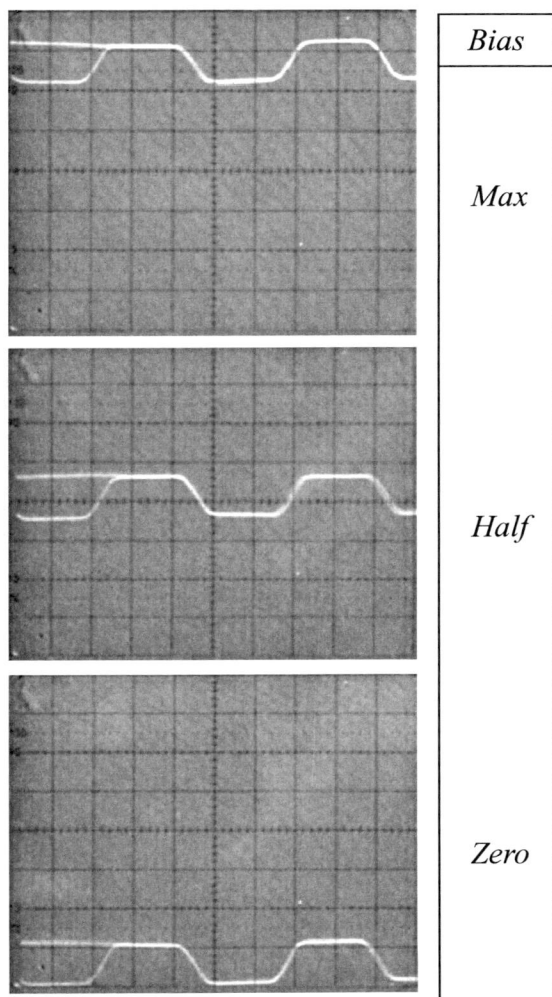

	Bias
	Max
	Half
	Zero

Figure 54

test source to subsequently free that electron from its valance bond. That would result in an increase in the test signal amplitude as displayed by the scope.

Experimental procedure:

1. Displayed test signal and bias signal voltages by the oscilloscope.

2. Set test source power supply so the displayed test signal was well below the display maximum.

3. Adjusted bias power supply from zero to a level that raised the scope to a level well above the original.

4. Observed the test signal voltage throughout the adjustment of step 3 above.

Results: As shown in Figure 54, the observed display of test signal voltage did not change over all levels of the bias source intensity. There was no additional detection capability created or removed by biasing the absorbing electrons. We can interpret this result to be a demonstration of the fact that electron absorption occurs in fixed increments of one or more quanta. In the light of Einstein's photo emission experiment, Section 4.3, it is probably one quanta per electron.

This is an "on-going" experiment with the objective of verifying results at different temperatures.

Quanata from the bias illumination undergo the same process. In this case the unmodulated bias voltage merely raised the entire trace level on the scope display.

The question addressed by Experiment #11 was: "Are there some quanta from the bias source which only partially energize valance electrons, but not freeing it?" If so, this would permit a portion of a quanta from the

4.5 Visible Wavelength H Field Detection Experiment # 12: Visible Wavelength H Field Detection

We have shown that the E and H fields of a radio frequency beam propagate independent of each other. And we have stated that essentially all visible wavelength detectors respond to light's E field only. These facts arouse one's curiosity. Can the two fields of light be separated? If they can, what would an H field world look like? The first step in a search for an answer to this question is to separate the E and H fields. This was done very easily in the radio frequency portion of the electromagnetic spectrum by use of individual "antennas," one for each type of field. However similar individual absorbers for light fields are not practical because of the small size of the wavelength, on the order of ten micro-inch.

A number of possible magnetic field detection devices were examined. None were able to accommodate the very high frequency of light, on the order of 10^{15} cps. But we found that it is possible to separate the two fields by taking advantage of differences in their indices of refraction for transparent material. Typical indices are listed in Table 3 below.

Experiment 12, which made use of the differing indices, consisted of five steps:

1. Generated a modulated white light.

2. Separated the E and H fields of the light.

3. Converted the H field to an H field with a "new" E field, via Maxwell curl, equation (6c).

4. Detected the "new" E field using a PIN detector.

5. Displayed the resulting electric signal.

Figure 55 is a sketch of Experiment 12's component arrangement. In step 1, the experiment started with a common automobile headlight lamp. The radiating light was modulated by a mechanical chopper to make it easier to distinguish the desired beam in the presence of interfering stray light.

In step 2, the modulated light passed through a beam-forming aperture and then through a glass prism. The type of prism is not critical. The difference between the E field indices of refraction for glass and air caused the beam to be refracted, or bent, upward by the prism. As shown by Table 3, the difference between the H field indices of refraction for air and glass is negligible. Therefore, the H field beam passed straight through the prism.

Table 3
Indicies of Refraction for Transparent Materials

Material	Index for E field	Index for H field
Air	1.000 3	1.000 000 37
Glass	1.4 to 1.6	1.005 01
Water	1.33	1.000 008 8
Plastic	1.49	unknown
Aluminum	conductor	1.000 023

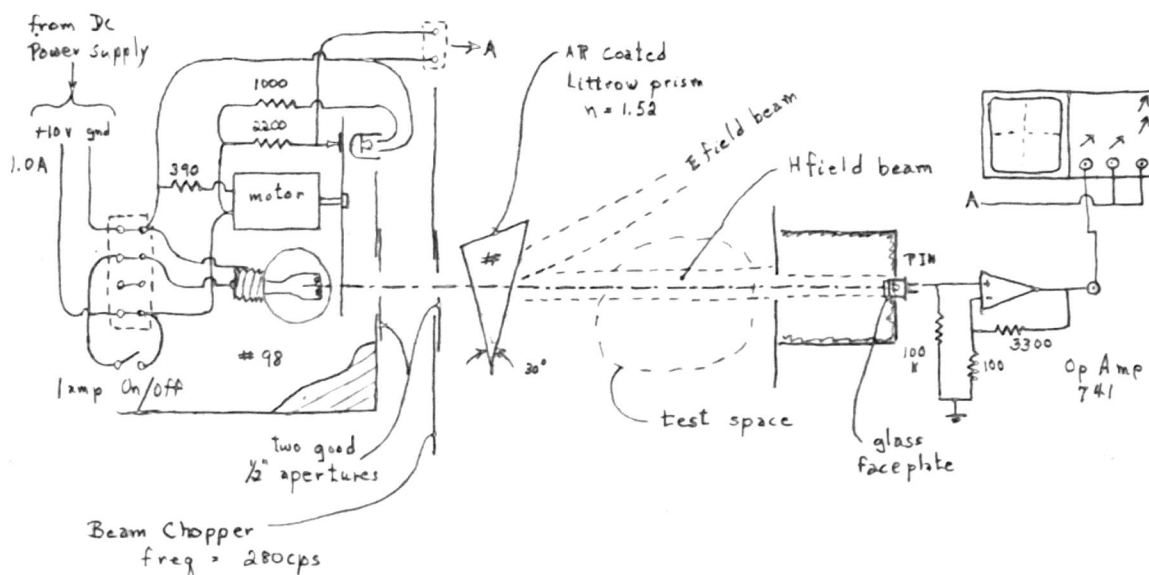

Figure 55

In step 3, as the H beam passed through the prism, there were sufficient free electrons in the glass. This new E field contained all of the modulation data present in the original H field beam. Overall efficiency of our experimental model was rather poor, approximately 1.0%. Experiments are continuing in search of a more efficient curl convertor. Radio frequency experiments, done with electron rich materials, demonstrated curl H and curl E efficiencies close to 100%.

In step 4, the PIN detector, Figure 58, responded to the new E field portion of the composite H plus E beam. It generated an electronic signal proportional to the magnetic part of the input beam. This step may use any

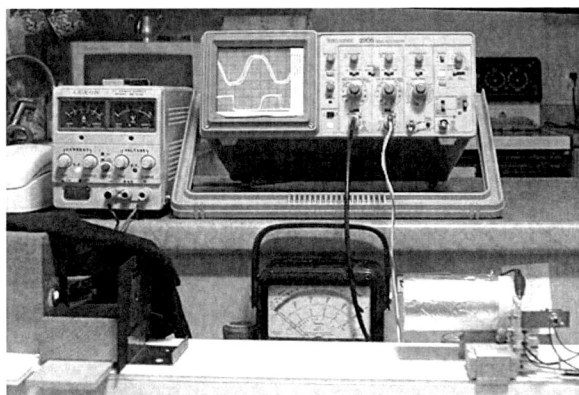

Figure 56

Figure 57

Figure 58

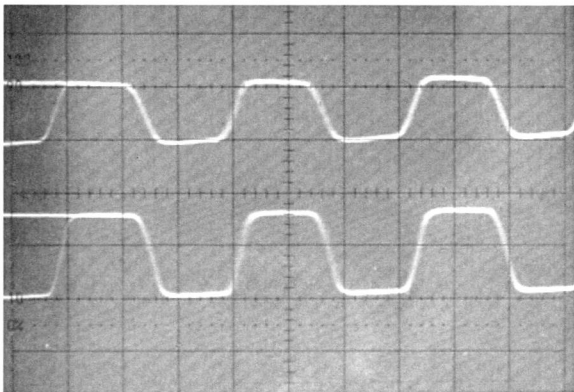
Figure 59

photodetector compatible with the desired wavelength band.

In step 5 the detected signal created from the H field beam was displayed by an oscilloscope, top trace in Figure 59. The E field is displayed by the bottom trace.

A number of tests were performed to assure ourselves that the isolated H field beam was indeed a beam of electromagnetic energy. The most convincing experiment was a measure of it's propagation coefficient, as defined by equation (14) in Section 2.2. We have determined

that propagating electromagnetic energy has a propagation dispersion proportional to $1/r^2$, where r = distance. We have also determined that all other autonomous propagating magnetic fields disperse intensity by $1/r^3$. (ref 29) The intensity of the H field beam emanating from the prism was found to disperse by $1/r^2$. Figure 60 on a following page is an example graph from this experiment. This is an "ongoing" experiment with the objective of increasing H $\Rightarrow$ curl E efficiency to 100%.

4.6 Scintillation

Experiment #13: Scintillation

Another fond memory from the author's younger years was of hockey practice at college in the hills of north-western Massachusetts. They had no formal rink and practiced on a large pond late in the evening when ice was hard and fast. The author enjoyed the mountains and winter sky more than the game. He was just meat for older players to practice on. On the other hand the night air was always cold and always crystal clear. He often wondered if a single light on the far-off hillside or perhaps one of the many stars was blinking a secret message for him.

Not so, the air played games with those light beams. High altitude winds were tossing to and fro random bubbles of air at slightly different temperatures, the same effect that causes clear air turbulence. These bubbles also bend starlight rays to make it appear that the stars are dancing. This is called scintillation. Astronomers have labored for centuries to overcome the resolution limits imposed by

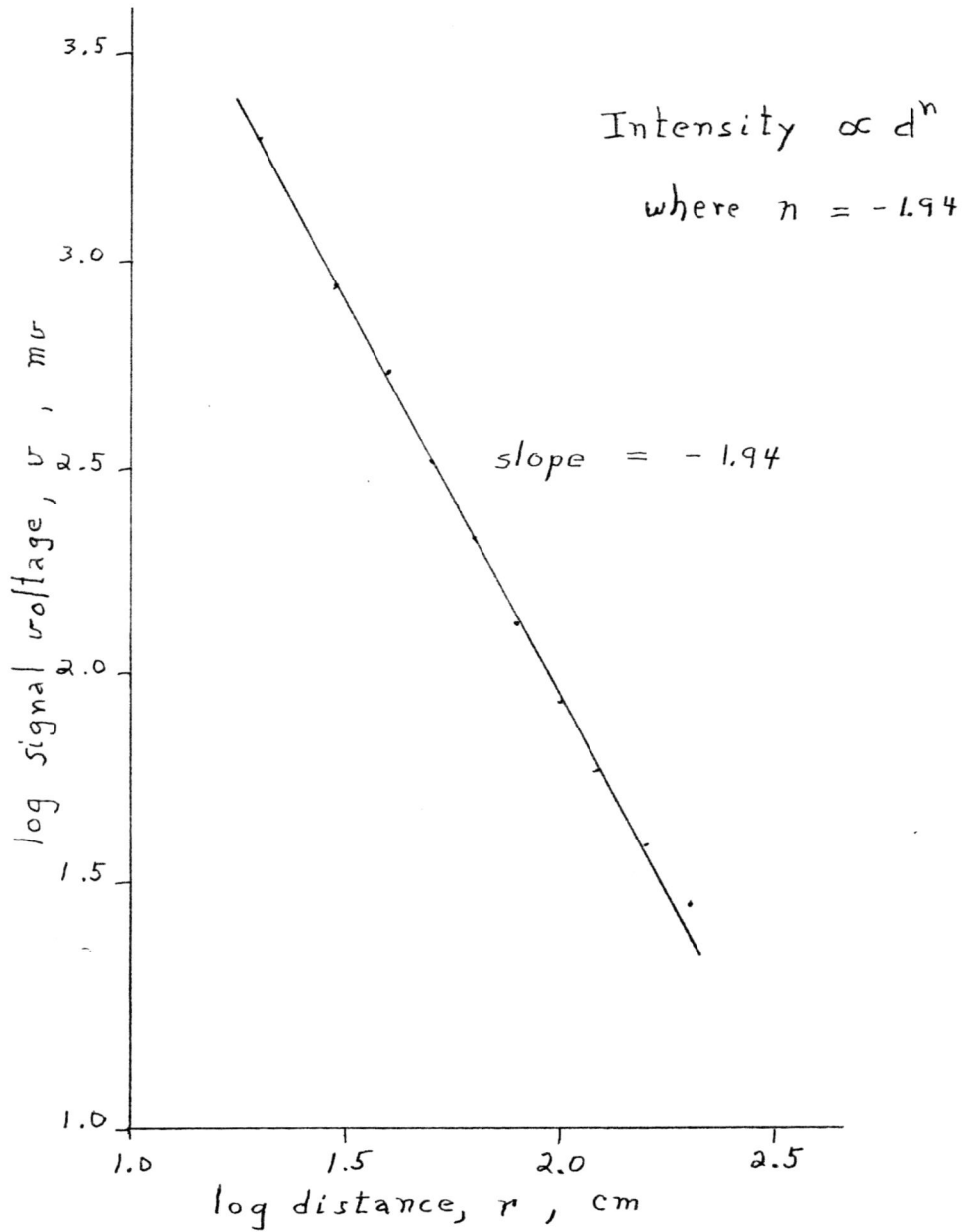

Figure 60

scintillation on their ground-based telescopes. Overcoming scintillation was one of the reasons for placing the Hubble Telescope in space beyond the atmosphere.

Scintillation is actually small but random angular changes in a light beam's direction. When photographing a star this results in random displacements of a star's image at the focal plane. An example of this effect is shown in Figure 61 below, which is a magnified photograph of a point source of light located approximately 200 feet away taken at dusk. The line of sight passed over a macadam parking lot that had been heated by the sun during the day. After sunset that heat rose up into the cool evening air causing random fluctuations in the air's index of refraction, which in turn caused random fluctuation in the E field's direction of propagation.

Experiment 7 has shown that radio-frequency E and H fields propagate essentially independent of each other.

The objective of Experiment 14 was to determine if light-wavelength E and H fields also propagate independent of each other. This was done by passing a vertical light beam through a tank of water. The water was

Figure 62

slightly disturbed to make gentle waves. Wave slopes then refracted the beam back and forth. This created bright and dark bands across the tank bottom. Figure 62 is a photograph of a similar situation in a sunlit swimming pool. It should be noted that the refraction occurs because of a difference between the E field index of refraction for water, 1.333, and air, 1.0003.

In Experiment 14 a small aperture below the tank bottom sampled the moving pattern and passed the light to a detector assembly identical to that described in Section 4.5. In this assembly the light was separated into two beams which were then directed to two detectors. The resulting signals were fed to channels 1 and 2 of an oscilloscope. Figure 63a-c are photographs of the oscilloscope traces. In all of the figures, the H field trace is above the E field trace.

Figure 61

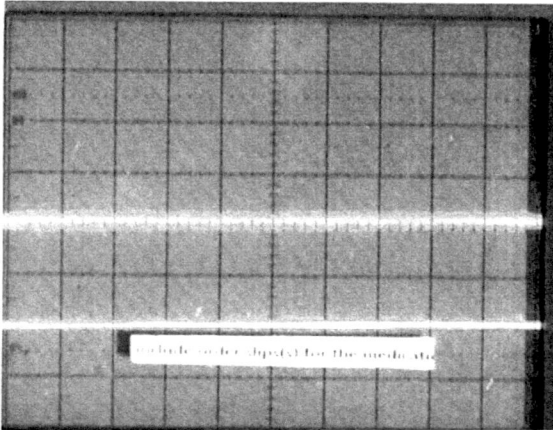

Figure 63a.

Detector Output Signals:

Reading taken with the shutter closed, i.e. no light entering the detector assembly.

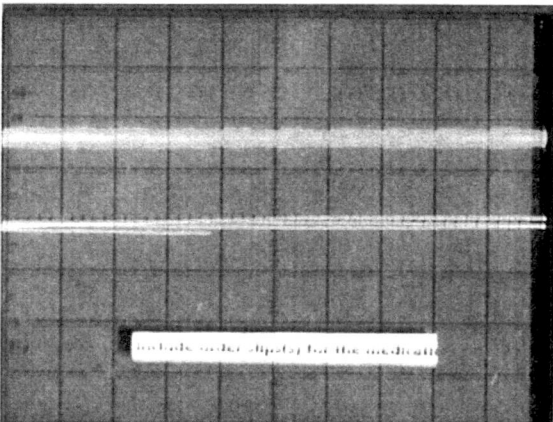

Figure 63b.

Reading taken with the shutter open, but no waves.

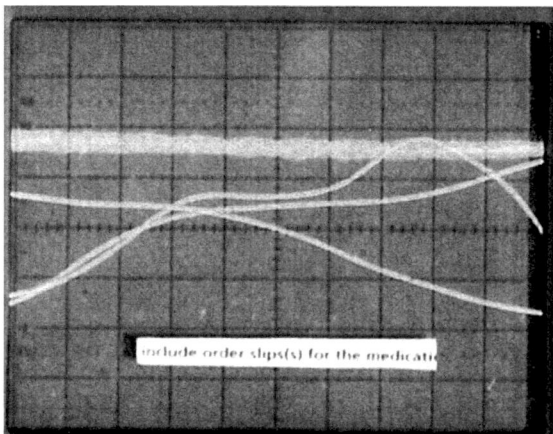

Figure 63c.

Reading taken with the shutter open, with waves.

Note in Figure 63c that the bottom trace, containing the E field signal, fluctuates significantly, while the top trace containing the H field signal has no intensity variation. Figure 64 is a repeat of the same conditions as in Figure 63c. Thus it can be said that the E and H field beams do indeed propagate independent of each other.

The reason the H field is not refracted by the surface wave slopes is due to the fact that the H field index of refraction is essentially 1.0 for both air and water, see Section 4.5. For those repeating this experiment be aware that it is easy to generate waves sufficiently large enough to refract E field energy into the H field detector.

Results from Experiment 14 imply that the angle-of-arrival of the H field portion of starlight would not be scintillated by those moisture-air or pressure-temperature fluctuations which cause image smear in ground-based telescopes.

Experiment #14 is an "on-going" experiment with the objective of adapting an imaging H field detector at the image part of a conventional telescope. The practicality of such an adaptation will depend upon successful completion of experiment 12.

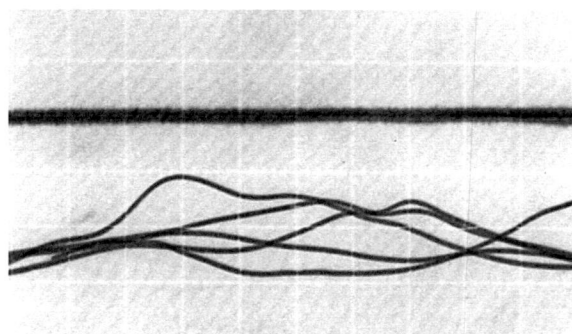

Figure 64

"Well I must have been wrong and you must be right, because you did an experiment and proved it."

- Ed Purcell's high-school teacher

POSTSCRIPT

It is hoped that you, a young student, have enjoyed reading about tweaking the physics establishment as much as the author has enjoyed experimenting for and writing this book.

As stated in the introduction, the purpose of this book is to stir the interests of college students in the study of physics and to show that learning Nature's Laws can be fun. It was also the author's intent to demonstrate that, with a little bit of thought and planning, meaningful experiments can be performed by an individual using simple materials and equipment.

To further these objectives, the author urges technically adventurous students among the readers to pick and explore one of those experiments described as "on-going." The author will provide drawings, photographs, hard to find materials (within reason) and personal suggestions on how to go forward. We would be eager to report resulting success or failures, with due recognition of the experimenters on our website: http://www.thewondersoflight.com.

> *"Fifty years of conscious brooding have brought me no closer to answering... What are light quanta? Of course every rascal thinks he knows the answer, but he is deluding himself."*
>
> *- Einstein, 1951* (ref 1)

Become an Einstein rascal!

Robert Crane

APPENDIX

A. Cover Photographs

a. Fringe pattern showing that interference is a lossless event

b. Recording of radio E and H field time separation

c. Pin-hole image from a relatively thin 30" diameter primary mirror after and before stress correction

d. Photographic evidence which suggests that light is a series of alternating E fields consisting of a random sequence of energy impulses, or quanta

e. Interference zone plate

f. Transfer function of the interferometer in the Hubble Telescope Fine Guidance Sensor

g. Sequence of pin-hole images versus exposure time from a six-segment-aperture optical system, Figure 66

h. A home-made spectrum

i. Absorption spectrum of heroin processing effluent

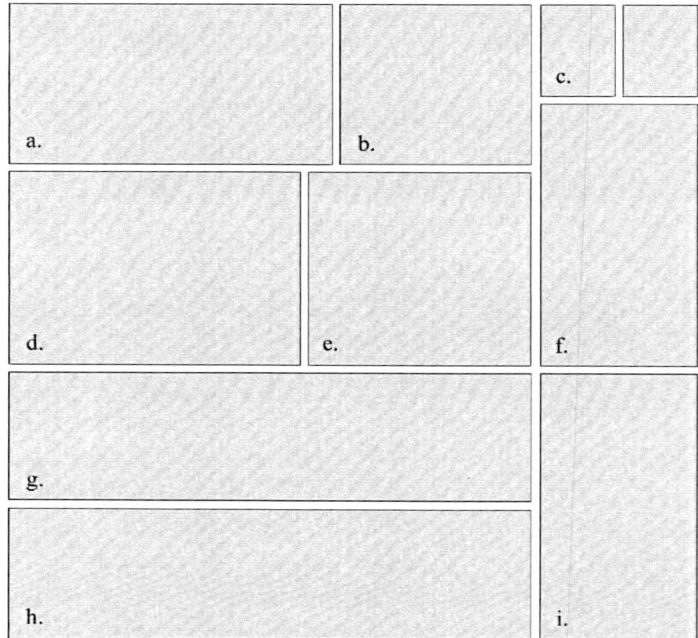

Figure 65

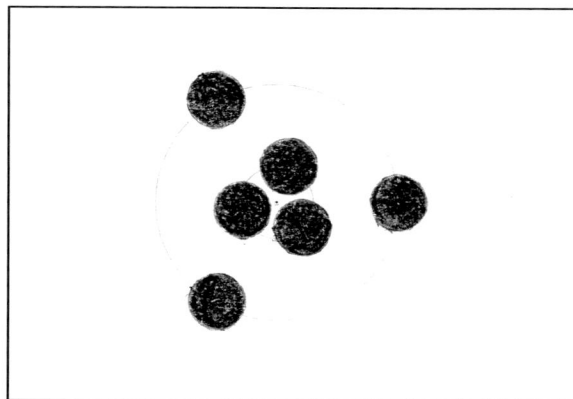

Figure 66

B. Law of Propagation

$$dispersion \propto (distance)^n$$

Source Dependence

Point E   Line   Plate

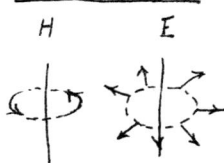
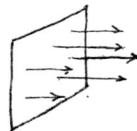

$$n = -2 \qquad n = -1 \qquad n = 0$$

Dual Source   ref (29)

magnet   Electric

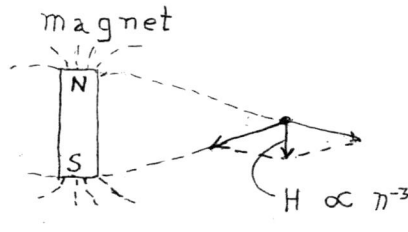
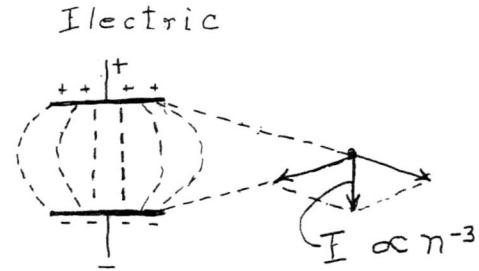

Point E Source

field strength $E = \dfrac{q}{4\pi\epsilon_0} \times \dfrac{1}{r^2}$ (volt)

field potential $V = \dfrac{q}{4\pi\epsilon_0} \times \dfrac{1}{r}$ $\left(\dfrac{volt}{m}\right)$

Electromagnetic

field strength E and $H \propto$ distance^{-2}

Collimated Beam

divergence $\propto \dfrac{\lambda}{D} \times$ distance

Figure 67

C. References

1 Zajonc, Arthur. Catching the Light: The Entwined History of Light and Mind. New York: Bantam Books, 1993.

2 Feynman, Richard P., et al. The Feynman Lectures on Physics: Mainly Electromagnetism and Matter. Reading, MA: Addison-Wesley Publishing Company, 1964.

3 Serway, Raymond A.. Physics: For Scientists and Engineers. Philadelphia: Saunders College Publishing, 1982.

4 Winch, Ralph P.. Electricity and Magnetism. Englewood Cliffs, NJ: Prentice-Hall, Inc., 1963.

5 Newman, James R.. Project Physics Reader 4: Light and Electromagnetism. Boston: Harvard Project Physics, 1967. Supported by: Carnegie Corporation, The Ford Foudation, NATIONAL SCIENCE FOUNDATION and Harvard University.

6 McGraw-Hill Encyclopedia of Science & Engineering, 1997 ed.

7 Hecht, Eugene. Optics, Second Edition. Reading, MA: Addison-Wesley Publishing Company, 1987.

8 Strong, John. Procedures in Experimental Physics. New York: Prentice Hall Inc, 1945.

9 Benenson, W., Harris, et al. Handbook of Physics. New York, Springer, 2002.

10 Newman 69-81.

11 Newman 145.

12 Newman 77 and 125. Serway 582.

13 Newman 146-150.

14 Newman 145.

15 Goldsmith, Oliver. An History of the Earth and Animated Nature, Volume. 8. England: J Nourse in the Strand, 1779.

16 Serway 859-60.

17 Millikan, Robert Andrews. The Electron: It's Isolation and Measurement and the Determination of some of it's Properties. Chicago: The University of Chicago Press, 1917.

18 Chaucer, Geoffrey. Canterbury Tales: The Miller's Tale, Circa 1387. London: Penguin Books, 1951.

19 Friedrich, Breitslav and Dudley Herschbach. "Space Quantization: Otto Stern's Lucky Star." Daedalus. Winter 1998.

20 Brown, R. Hanbury and RQ Twiss. "Selected Papers on Long Baseline Stellar Interferometry." Nature 1956: pgs 27-29.

21 Saunders, J.B. "Construction of a Koesters Double-Image Prism." Journal of the National Bureau of Standards, Jan, 1957.

22 Photo Credit: Hester, Jeff and Paul Scowen.(Arizona State University) and NASA. NASA's John F. Kennedy Space Center and Goddard Space Flight Center. http://www.nasa.gov/missions/deepspace/f_two-hubbles.html

23 Patent Number 4,735,114. May 10, 1988. Fabry-Perot Scanning and Nutating Imaging Coherent Radiometer. Inventor: Crane, Robert Jr., and Dunavan, David. Assigner: The Perkin-Elmer Corporation, Norwalk, CT.

24 Challoner, Jack. The Visual Dictionary of Physics: Radio Waves. London: Dorling Kindersley, 1995: pg 38, permission requested.

25 Hecht 456.

26 Townes, Charles H. How The Laser Happened: Adventures of a Scientist. Oxford: Oxford University Press, Inc., 1999.

27 Thomas Young, "Experiments and Calculations Relative to Physical Optics," 1855: found in Ref 5, pgs 15-27.

28 Taylor, Goeffrey. "Interference Fringes with Feeble Light." University of Cambridge, Cambridge Philosophical Society 15, 1909: pg 114.

29 Serway 411, the geometry of a two-pole source which leads to a propogation law co-efficient of n = -3.